NUCLEAR
Neighbourhoods

COMMUNITY RESPONSES
TO REACTOR SITING

NUCLEAR
Neighbourhoods

COMMUNITY RESPONSES
TO REACTOR SITING

J. Richard Eiser
Joop van der Pligt
Russell Spears

UNIVERSITY
of
EXETER
PRESS

First published in 1995 by
University of Exeter Press
Reed Hall, Streatham Drive
Exeter, Devon EX4 4QR
UK

British Library Cataloguing in Publication Data
A catalogue record of this book is available
from the British Library

Hardback   ISBN 0 85989 455 X
Paperback   ISBN 0 85989 456 8

Printed and bound in Great Britain
by Short Run Press Ltd, Exeter

Contents

Preface

The future of nuclear power has been the most controversial issue in the field of energy policy over the last twenty years. There are many reasons why this is so. Nuclear technology is complex, yet the end-point of this technology is simple enough: the production of extreme heat to produce steam to turn turbo-alternators to generate electricity. This end-point can be achieved by other, much cruder, technologies. The hope was once that, by employing sophisticated science to 'harness the power of the atom', we would have a state-of-the-art technology that would be, almost by definition, more clever, clean and efficient. These hopes are still held by some, but more modestly now. The cleverness of nuclear technology comes at a price. The capital costs of building nuclear power stations have been consistently underestimated. The 'back-end' costs of decommissioning defunct reactors and disposing of radioactive waste seem to have been almost completely disregarded in the early days of nuclear expansion. Then there is the fact that accidents can happen, albeit rarely, but with such worrying and widespread consequences that any accident is one too many.

Yet this is far more than a merely technical debate between different engineers and economists. Public attitudes, and particularly those of local residents, are an inescapable part of any calculation. If any nuclear development is to proceed, the industry faces a difficult task in trying to reassure local communities that their interests will be protected. Even where such reassurance is reasonably convincing, the process can take a long time. If the plans are the subject of a public inquiry, it can take a very long time indeed, further adding to the industry's costs and casting further doubt on the profitability of the development. The intensely political nature of the debate is under-lined by the fact that the perceived needs of local communities and of national governments may diverge. Even in the age of nuclear disarmament rather than the nuclear arms race, the possession of nuclear expertise has implications for national security. Strategic planning for the country's future energy needs—as well as the need to reduce emissions of 'greenhouse' gases from fossil fuels—may imply a need to preserve a mix of technologies.

Shorter-term political considerations may also be influential. As this book goes to press, the British government has announced its intention to privatize the nuclear power industry during 1996. Various features of the plan seem designed to provide the Conservative government with sufficient revenue to allow it to make tax cuts in the run-up to the next general election. Notably, Nuclear Electric is to be merged with its more profitable neighbour, Scottish Nuclear, sacrificing the possible benefits of competition so as to create a single unit expected to be more attractive to investors at the time of sale. At the same time, while the proceeds of the privatization will be partly used to cover the costs of the decommissioning of older (Magnox) reactors, there are concerns that some of these costs may still need to be paid for out of tax revenue at a future date. Significantly, the government has set its face against providing subsidies, either from tax revenue or levies on the consumer, for the building of more nuclear power stations. If capital for the building of new reactors is to be found, it must be found in the market. Many analysts believe that this signals the end of expansion of the British nuclear power industry for the foreseeable future.

Our own research started at a time when a more assertive and state-controlled nuclear industry was talking confidently of constructing a new generation of Pressurized Water Reactors (PWRs) throughout Britain. It was claimed that these would provide an economical and safe source of power, without many of the political costs associated with dependence on coal or oil. One minor problem with this idea was finding sites for these new power stations that would be physically suitable and would not arouse excessive local opposition. Of course, there would always be some objections, but these could surely be overcome by proper consultation. After all, the obvious places to build any such stations would be in coastal regions away from major centres of population, so the number of local objectors would be relatively limited anyway.

Among the regions considered to be in need of a new nuclear station was the South-West of England, where the three of us all lived and worked at the time. As we shall see, matters did not turn out to be quite so simple here from the industry's point of view. For better or worse (and it is not for us to say which), the industry's plans ran into local opposition which was greater in extent and more varied in kind than perhaps had been anticipated. Our main aim in this book is therefore to draw together the findings from a series of public surveys we conducted as the industry's plans were announced and subsequently revised. These surveys describe the viewpoints of residents in a number of small communities, chosen for the most part because they might be affected by the industry's plans for expansion. The main theme of our work is that people have *reasons* for their opinions, whatever these opinions may be. Often in public debates one hears only contrasting opinions and not the reasons behind these, but if one is to understand why people adopt different points of view, it is vital for these reasons—all of them—to be heard. To have a different point of view on an issue really does mean seeing that issue differently and regarding different aspects and arguments as the most important.

Our thanks are due to the many people whose advice and assistance has been invaluable—most recently to Karin George, who typed this manuscript, to our colleagues at the Universities of Exeter and of Amsterdam who have sustained our interest in this field, and to the Economic and Social Research Council who provided the funding for the research (through project grant number D00250009). Most of all, however, we are indebted to the many ordinary people of small communities in the South-West of England who participated in our studies, responding to sometimes lengthy and complex questionnaires with care and patience. If this book can give voice to just a part of the subtle diversity of the views they expressed to us, we will have repaid something of what we owe them. This is their story.

Chapter 1

Nuclear Energy:
Public Opinion and Local Attitudes

From the late 1950s to the mid-1970s the future of nuclear energy seemed assured. In the late 1970s the public became more involved in nuclear energy issues and entered the once-exclusive domain of energy policy-making. This increased involvement is also reflected in the rapid growth of the environmental movement. Since then, public acceptance has become a crucial issue in the nuclear debate both at a national and local level. This chapter presents a brief overview of public reactions to nuclear energy. First we will describe public opinion data concerning nuclear energy in general and discuss the changes which took place over the past two decades. Next we will focus on *local* public opinion. Differences between general attitudes and attitudes towards the local building of a nuclear power station will be discussed, as well as the major factors determining the local perspective. Local reactions to nuclear power are the central theme of this book and in this chapter we provide a brief overview of the existing literature on this issue.

Public opinion and nuclear energy

Melber, Nealey, Hammersla and Rankin (1977) compiled the first comprehensive overview of survey research on the issue of nuclear energy. At that time (1976–1977) there were twenty-seven national US surveys of nuclear attitudes available. There was one each in 1960, 1970 and 1973, three surveys were conducted in 1974, eleven in 1975, and ten in 1976. This increase in survey frequency illustrates the rise of nuclear power as a controversial public issue. In Europe it took a bit longer before the frequency of public opinion polls increased; the first coordinated, large-scale EC surveys were carried out in the late 1970s and early 1980s.

The increased public interest is also illustrated by media coverage of the nuclear issue. Rankin and Nealey (1978) report a *fivefold* increase of US newspaper and magazine coverage between 1972 and 1976. Television's attention to the nuclear issue increased *ninefold* in the same period. In chapter 5 we will present a more detailed analysis of media attention to the nuclear energy issue. Nealey, Melber and Rankin (1983) present an extensive overview of US public-opinion data collected between 1975 and 1981. These surveys show a steady increase of the number of opponents from the mid-1970s onward. This increase accelerated as a consequence of the Three Mile Island accident, mainly at the expense of the 'undecided' category. Overall, the number of opponents and proponents was about equal, with a small majority of men being in favour and a small majority of women being against nuclear energy. Since the late 1970s a majority of the public opposes the *local* building of a nuclear power station.

A number of other conclusions can be drawn from these surveys. The first is that the accident at Three Mile Island (TMI) in 1979 had a significant impact on public attitudes towards the construction of additional nuclear power plants. After TMI there was a substantial decline in public acceptability of nuclear power. Although there has been some variation in public acceptability in the years following the accident, it did not return to its previous (pre-TMI) level. US surveys carried out in the early 1980s, confirm that public attitudes towards building more nuclear power plants are more negative than those reflected by the survey data prior to 1979 (see Nealey, Melber and Rankin, 1983). The years immediately following TMI thus show a significant increase of opposition as compared to pre-TMI levels and no significant trends for acceptability to return to its old level.

Other survey data also indicate that the accident at Three Mile Island (together with Chernobyl probably the most widely known event in the history of nuclear power) had a profound impact on public opinion. For instance, US opinion polls at the time showed that more than 95 per cent of the public had heard or read about the TMI accident, and 50 to 70 per cent believed such an accident could happen again. Canadian surveys (Decima, 1987) are consistent with US survey data, with nearly 80 per cent of the total population

2

opposing the building of more nuclear power stations. Moreover, Canadian surveys show that even those in favour of the present use of nuclear energy tend to oppose the building of more nuclear power stations (nearly 70 per cent). Figure 1.1 provides an overview of public attitudes in the United States in the period 1975–1988. It illustrates the effects of the TMI accident.

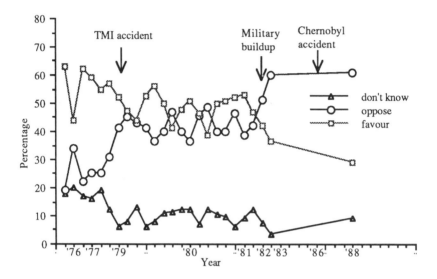

Figure 1.1   Public attitudes towards the building of more nuclear power plants in the United States.

Note: Adapted from: Rosa and Freudenburg, 1993, p. 48.

TMI was followed by a period of 2 years in which support and opposition to nuclear power seesawed. On average, support did not return to pre-TMI levels and the undecided category increased slightly. In late 1981 opposition surged upward again when the USA increased its defence spending under president Reagan and introduced expensive programmes such as the 'Star War' plans. This increase remained stable until the Chernobyl accident in 1986. The accident at Chernobyl further reduced the number of supporters and increased the number of people who are undecided. As we will see later, Chernobyl had only a modest effect on public attitudes in the

3

United States as compared to public reactions in Europe. Mazur (1990) has argued that the increase in public opposition in the United States in the early 1980s was mainly due to public concerns about the increased expenditure on nuclear arms as announced by President Ronald Reagan immediately after he assumed office. In the context of the present book it is also interesting to look at the development of attitudes towards the construction of more nuclear power plants.

Rosa and Freudenburg (1993) showed that nuclear attitudes seem consistently more negative than attitudes towards other energy sources. Nuclear energy is clearly less popular than other energy supply technologies (coal, natural gas and oil). They refer to a survey in 1985 asking respondents to name the energy source most dangerous to human life. A total of 82 per cent mentioned nuclear energy, while coal, natural gas and oil were mentioned by respectively 4 per cent, 5 per cent and 1 per cent of the respondents. Similarly, when asked to select the energy source least acceptable for large-scale use nearly 60 per cent chose nuclear power, 20 per cent chose coal, 4 per cent natural gas and 9 per cent oil (see also Roper, 1985). In later chapters we will present the opinion of samples in the South-West of England on the above issues.

More recent US opinion polls show less extreme opinions than in the late 1970s and early 1980s but still indicate a majority of the public opposing the construction of more nuclear power plants. A survey carried out in 1991 for *Time* magazine and CNN (Cable News Network) revealed that 52 per cent of the US public opposed the construction of more nuclear power plants, 40 per cent favoured more plants and 8 per cent were unsure (Atom, 1991).

The first standardized cross-national opinion polls in the European Community were conducted in the late 1970s and early 1980s. The most elaborate survey took place in 1982 (Commission of the EC, 1982). Between March and May of 1982 an identical set of twenty questions was presented to large representative samples of the population in ten countries of the European Community. Figure 1.2 presents an overview of opinion shifts between 1978 and 1982 and shows considerable variations within the EC.

France (the country in the EC with the highest number of nuclear power plants) is the country with the most favourable public opinion. In 1982, a small majority of the French population thought the

4

development of nuclear power worthwhile. German public opinion also moved in a direction of higher acceptability, but because of the large percentage of those undecided, supporters are not in a majority (see Figure 1.2 (a). The low level of public acceptability in Germany is also reflected in local and federal opposition to expansion of the nuclear industry. For instance the (recently built) nuclear reprocessing plant at Kalkar (a joint initiative by Germany, Belgium and the Netherlands) will *not* become operational due to federal and local opposition. Later in this chapter we will return to the issue of local reactions to nuclear power stations. As can be seen from Figure 1.2 (b), a number of EC countries have seen a steady increase of opposition. This increase usually took place at the expense of the undecided category. The overall percentage of support remained stable at about 37 per cent.

These findings from 1978–1982 show a public that is divided about the issue. It should be noted that about 25 per cent of the European citizens had either no opinion or found the issue not particularly interesting. Further breakdowns of these figures according to political preference revealed that the political left–right dimension is clearly related to views about nuclear development. A majority (52 per cent) of those on the right of this political dimension thought the development was worthwhile as compared to just under 30 per cent for those on the left.

One of the issues discussed in the previous paragraphs concerns the impact of major accidents on public opinion. The impact of the TMI accident is clearly illustrated by US and Canadian data showing a significant increase in public opposition. About seven years after the TMI accident a much more serious accident took place at Chernobyl in the Ukraine. The reactor accident at Chernobyl on April 25–26 1986 resulted in a serious threat of radioactive contamination to various countries in Northern, Central and Western Europe. Extensive opinion surveys were carried out after the accident both in Europe and in the US.

As in the aftermath of the TMI accident, support for nuclear power declined in most countries. Some recovery was noticeable after a period of time, but generally, this recovery was not strong enough for public opinion to return to pre-Chernobyl levels.

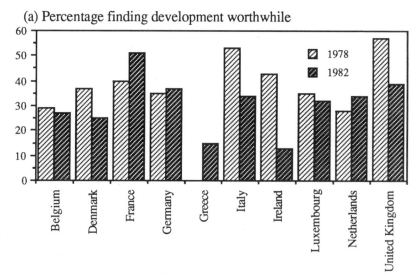

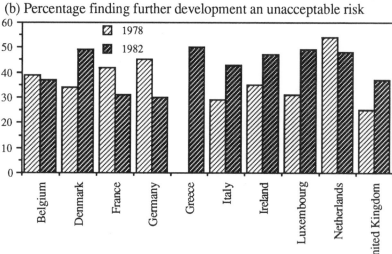

Figure 1.2 Public opinion about nuclear energy in the EC 1978–1982

Note: Greece was not included in the 1978 survey
Adapted from Commission of the EC (1982)

Increases in public opposition were most pronounced in Finland, Yugoslavia and Greece (over 30 per cent), and were substantial in

Austria, West Germany and Italy (over 20 per cent). More moderate changes took place in the United Kingdom, the Netherlands, France, Sweden and Spain (12–18 percent). Figure 1.4 presents an overview of these changes.

The US was not affected by the fallout, but public opposition to nuclear power increased by five per cent to reach a peak of nearly 50 per cent, the highest ever reported (*Newsweek*, 1986). More dramatic changes can be seen in attitudes towards the building of a new nuclear power station locally. Opposition to the building of a nuclear power station in the locality was already quite substantial after the TMI accident and showed a further increase (70 per cent being opposed to the building of a local nuclear power station).

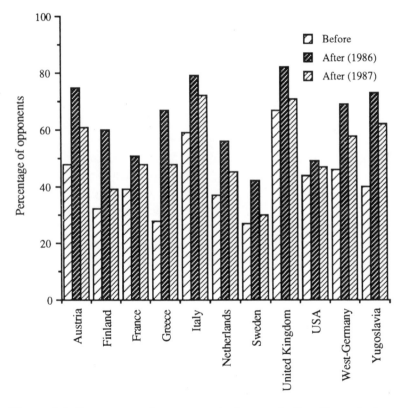

Figure 1.3 Percentage of opponents before, and after Chernobyl.
Note: Adapted from Van der Pligt, 1992

The level of opposition in the US more or less returned to pre-Chernobyl levels a year later (Renn, 1990). As shown in Figure 1.3. this was not the case in Europe, where the consequences of the accident were more severe than in the US.

In Canada, over a year after Chernobyl, 79 per cent of the population disagreed with the statement that 'there is really no chance that there will be a major accident in Canada's nuclear power reactors like the accident at Chernobyl' (Decima, 1987).

Figure 1.3 summarizes the increases of opposition immediately after the accident and gives an indication of the stability of these changes (survey data collected at least a year after the accident). In all countries there is some reduction of public opposition a year after the accident. However, in most countries, this return to pre-accident levels was incomplete. A year after the accident public opposition was still between 5 and 25 per cent higher than before the accident. Although no public opinion data are available from central and eastern European countries, observers reported a growing opposition to nuclear power in Poland, Hungary and Czechoslovakia (see for example *Nucleonic Week*, 1986; and Renn, 1990).

In Chapter 6 we will return to the issue of nuclear accidents and their effects on public attitudes towards nuclear energy. In that chapter we will discuss the effects of publicity concerning a series of minor incidents and radioactive emissions from the Sellafield reprocessing plant in North-West England. We will also present data on the effects on public attitudes of the Chernobyl accident. In both cases our major concern will be how people generalize from incidents and accidents elsewhere to make judgments concerning the desirability of nuclear developments closer to home.

Local opinion

Generally, local opinion about the possible (local) development of a nuclear power plant or a nuclear waste repository is more extreme and more 'anti' than opinions about nuclear energy in general. It seems that the siting of nuclear facilities has become almost a textbook illustration of 'locally undesirable land uses', or LULUs (Popper, 1981). Surprisingly, even the idea of a local nuclear power plant remained relatively uncontroversial until the mid-1970s. Later, most opinion polls show a clear majority of people opposing the building of a nuclear power station in their locality. As we mentioned before, the TMI accident had a clear impact on local public opinion in the US. Immediately after TMI over 60 per cent of the US public opposed the *local* construction of a nuclear power plant; an even higher percentage was found immediately after the Chernobyl accident. Recent opinion polls still show a majority of the US public opposing the local construction of a nuclear power station. Canadian surveys confirm this 'Not–In–My–BackYard' or NIMBY phenomenon. For instance, 58 per cent of the New Brunswick population regarded the building of a nuclear power plant in their province as a major threat and 65 per cent of the residents of Saskatchewan opposed the building of a nuclear power station in their province (Decima, 1987).

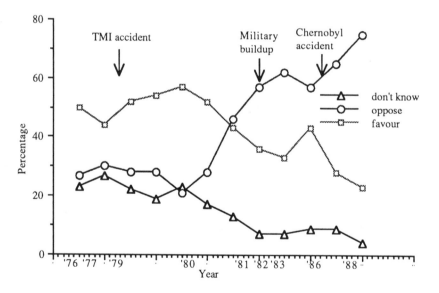

Figure 1.4   Public attitudes towards the construction of local nuclear power stations.

Note: Adapted from Rosa and Freudenburg, 1993, p. 57.

9

Figure 1.4 provides an overview of public attitudes towards the construction of local nuclear power plants in the United States. In the early 1970s there was considerable support (over 50 per cent) for the building of a local plant. On average the level of support was twice the level of opposition. Opposition to the local construction began to increase from 1976 onwards, accelerated after TMI and reached a peak after the Chernobyl accident, when more than 70 per cent of the public opposed a local nuclear power plant.

Another indication of the low acceptability of local nuclear power stations is given by the 'tolerance distance' for different types of facilities. In this research respondents are generally asked to indicate how close each facility could be built to their homes 'before you would want to move to another place or to actively protest'. Rosa and Freudenburg (1993) describe that even the disposal site for toxic chemical waste was more popular (81–84 miles) than nuclear power. The latter has a mean tolerance distance of 91 miles (see also Mitchell, 1980).

The NIMBY phenomenon is also clearly reflected in public opinion data concerning nuclear waste disposal. Generally, a majority of the public opposes the local construction of a waste repository. Moreover, of those opposed, more than 30 per cent indicated they would be more likely to support waste disposal if the site selected was not in their state (Nealey, Melber and Rankin, 1983). In the same study 63 per cent of the interviewed politicians would be more likely to support waste disposal initiatives if not located in their state. Waste disposal issues play an important role in public opinion. For instance, when people are asked to make a direct comparison between waste management and reactor safety, the majority believes the former to be a bigger problem (Nealey *et al.*, 1983). Recent Canadian opinion polls show a similar pattern; a total of 56 per cent indicated nuclear waste disposal to be of greater concern than the potential of a nuclear reactor accident (41 per cent). Not surprisingly, nearly 70 per cent of local residents opposed a nuclear waste repository in Northern Ontario (Decima, 1987).

European surveys confirm the above findings. For instance, in the 1982 survey respondents were presented with various hazards associated with nuclear energy, and were asked to indicate the hazards that worried them most. A total of 57 per cent selected the

10

nuclear waste issue, 51 per cent selected routine radioactive emissions from power stations, and 23 per cent mentioned a major accident (explosion) of a nuclear power station. Recent polls conducted in the UK confirm the importance of the nuclear waste issue. When confronted with a variety of environmental hazards the number one issue was nuclear waste dumping. A total of 64 per cent of the respondents was worried about this issue (Van der Pligt, 1992).

A number of surveys have either compared level of acceptance of a nuclear power station among people who live near one with that of people who do not, or monitored public opinion in a locality during the planning and construction stages of a nuclear power station. Overall, there is mixed support for the idea that familiarity leads to greater acceptance of nuclear facilities. Melber *et al.* (1977) mention eight studies which monitored local acceptance of a nuclear power plant during construction. Only two of these showed a significant increase in acceptance over time, while one locality showed a significant increase in the level of opposition. The 1982 EC survey suggests a greater acceptance of nuclear energy in countries which have a higher number of nuclear power stations. Comparisons of those within a 30 or 60 mile radius with the remaining respondents, however, revealed only marginal differences. Other research on local opinion also fails to support the notion that familiarity leads to more support for nuclear facilities (Warren, 1981). In Chapter 3 we will discuss the effects of familiarity in more detail. In the next section we will briefly discuss research on local attitudes and the perception of costs and benefits that determine local reactions.

Local attitudes towards nuclear power stations

In the previous sections we gave an overview of the results of large-scale surveys conducted in the US, Canada and the European Community. The major benefit of these surveys is their longitudinal character, enabling us to look at changes over time, for instance in the context of nuclear accidents. A shortcoming of this research is that opinion polls are usually limited to a handful of questions. This has two drawbacks. The first concerns the fact that questions aimed at tapping people's opinion about an issue can be framed in a variety of ways. In the area of nuclear energy, Thomas and Baillie (1982) found at least six different questions which they suggest could tap different aspects of the issue at hand, and as a consequence, lead to different answers. The meaningfulness of answers will, to a large extent, depend on the way the questions are framed, but a restricted set of questions allows little opportunity to correct for such biases.

The second drawback concerns the limited amount of information about the structure of public attitudes that is provided by most large-scale survey research. Opinions are usually based on more specific beliefs about risks and benefits. Most surveys are, however, limited to a few questions, making it difficult to study attitudinal structure and process in more detail. In the next two sections we focus on research of *local* attitudes towards nuclear facilities during their construction and during the process of siting nuclear power stations. The more detailed data from these studies allow a more structural description of attitudes to be attempted.

Local attitudes during construction

In this section we will concentrate upon local acceptance of nuclear power plants. Most research in this area employs expectancy-value models to examine the structure of attitudes towards a specific nuclear facility (see also Chapter 3). As indicated in the previous sections, local opinion about the building of a nuclear power station is often more pronounced than the opinion about nuclear energy in general. This is likely to be related to the perceived risks and benefits to the affected communities. One of the first studies investigating local acceptance of a nuclear plant covered a period of five years, and followed local acceptance throughout the construction of the power station (Hughey, Sundstrom and Lounsbury, 1985). The study took place in Trousdale County in central Tennessee, which was selected to host one of the world's largest nuclear generating power plants. The Hartsville nuclear power plant was expected to occupy 1900 acres and would contain four reactors with an estimated capacity of 1250 megawatts each. The total costs were initially estimated at $2.5 billion. Estimated costs increased dramatically, however, and plans for two reactors were postponed soon after construction began (cf. Hughey *et al.*, 1985). Construction was also delayed. The entire project should have been completed by 1983. In fact, construction of the first power plant proceeded until late 1982 when Tennessee Valley Authority announced plans to (indefinitely) delay further construction. At this time, construction of the first of the four reactors was still unfinished. Construction of the first reactor required an average of 1100 workers; peak construction involved more than 5000 workers (which equals the total population of Trousdale County).

Prior to the start of the construction of the plant in 1975 a majority of the residents (69 per cent) indicated that they did not object to the construction of the plant. Five years later, this support was reduced to 45 per cent. In 1975 over 60 per cent of the residents would have allowed construction of the nuclear power plant within 20 miles of their homes. Five years later this was reduced to less than 35 per cent; a majority (54 per cent) wanted the nuclear power station at a distance of 100 miles or more. Results of their study thus indicated a substantial decline in local acceptance of the nuclear power plant during construction. It needs to be noted, however, that the design of the study made it difficult to establish a direct causal link between attitudinal shifts and changes in perceived risks and benefits. Both the economic decline in the late 1970s and the TMI accident could have played a part in the considerable reduction of local acceptance.

Hughey *et al.* (1985) clearly showed that local acceptance represents a trade-off between benefits, costs and risk. Favourable attitudes to the construction were related to the view that benefits such as more jobs and better schools were likely consequences of the construction and operation of the nuclear power station, while adverse effects such as health hazards and drugs in the schools were not very likely. Opponents, on the other hand, viewed health hazards and community disruption as more likely. They were also more pessimistic about potential benefits. After five years the perceived balance in costs, risks and benefits had changed towards seeing the plant primarily as a source of concern.

In analysing the assessment of costs and benefits, Hughey *et al.* (1985) factor-analysed the ratings of 27 possible positive and negative outcomes. Results revealed five factors. Table 1.1 provides more information about the five factors obtained in their analysis.

13

As can be seen in Table 1.1, the first factor (economic growth) included such possible consequences as increased local business, more jobs, and more stores and shopping areas. Factor II (hazards to safety and environment) included radiation hazards, air pollution, the pollution of a nearby lake, and sabotage. Factor III (general economic benefits) included industrial development and more entertainment. Factor IV (community disruption-1) included issues such as traffic congestion and drugs in the schools. Factor V (community disruption-2) included issues such as increased crime and crowded schools.

Table 1.1   Factors underlying attitudes towards a local nuclear power plant obtained by Hughey *et al.* (1985)

Factor I:	*Economic Growth*	Factor III:	*General Economic Benefits*
	Increased business		
	More jobs		More entertainment
	More stores and shopping areas		Cheap electricity
	Increased land values		Industrial development
	Better pay		Town becomes a
	Housing shortage		tourist attraction
	Meeting new people		Better schools
	More recreation areas		
	Recognition of town	Factor IV:	*Community Disruption-1*
	More billboards		
			Drugs in schools
Factor II:	*Hazards to Safety and Environment*		More taverns and bars
			Increased noise
			Traffic congestion
	Radiation hazards		
	Sabotage	Factor V:	*Community Disruption-2*
	Air pollution		
	Pollution of lake		Increased taxes
	More foggy days		Crowded schools
			Increased crime

Note: Adapted from Hughey *et al.* (1983), p. 665.

An important finding obtained by Hughey and his associates was that some potential outcomes were seen as more important than others; health hazards and environmental damage were apparently

14

more important than economic benefits and local disruptions. Sundstrom, Lounsbury, DeVault, and Peelle (1981) obtained similar findings.

Table 1.2   Multiple regression analysis of local acceptance of a nuclear power plant

Predictor	multiple R	multiple R^2	change in R^2
Hazards	0.52	0.27	0.27
Community disruption (1)	0.57	0.33	0.06
Economic growth	0.61	0.37	0.05
Community disruption (2)	0.62	0.38	0.01
General economic benefits	0.62	0.38	0.0

Note: Adapted from Hughey *et al.* (1983), p.665.

Both studies point to the role of relative importance or salience of specific possible consequences. In Chapter 3 we will discuss the issue of perceived salience in more detail.

Table 1.2 summarizes the findings obtained by Hughey *et al.* concerning the relative contribution of the five factors. Their findings showed that hazards to local safety and the local environment were the best predictor of local acceptance, followed by 'community disruption' and 'economic growth'. Only the first three factors contributed significantly to the overall prediction of acceptance.

The research discussed in this section therefore shows that attitudes do not necessarily become more favourable during construction. Moreover, the findings suggest that even after a long process of adaptation to the idea that a nuclear power station will be part of local life, a majority of local residents oppose such a development. Finally, the findings of Hughey *et al.* show that different factors determine the attitudes of opponents and supporters of the building of a nuclear power station in the locality. In the next section we return to this issue and will briefly describe a study of local attitudes during a siting controversy.

Local attitudes during siting

Woo and Castore (1980) examined attitudes during a nuclear power plant siting controversy. In this study the perceived benefits and costs were also closely related to attitudes towards the proposed nuclear power plant. All participants lived within 30 miles of the proposed site in northern Indiana, USA. In their analyses, Woo and Castore (1980) focused on differences in the perception of these costs and benefits as a function of attitude towards the nuclear power plant. Their results indicated that understanding the positions of three attitude groups (pronuclear, antinuclear and neutral) required two types of information. First, the beliefs about the possible consequences held by the attitude groups *and*, second, the subjective weights or importance attached to these beliefs.

Table 1.3 summarizes the frequency with which the various potential positive and negative outcomes were associated with the proposed nuclear power plant by the three attitude groups. People's perception of the positive and negative outcomes was significantly related to their overall evaluation of the proposed nuclear power plant.

The pro-group saw more benefits than costs; the reverse was true for the anti-group. Subjects were also asked to indicate the importance they attached to each of the possible consequences. Not surprisingly, the 'pro' groups attached much more value to possible economic benefits, while the 'anti' group was most concerned with potential health and safety issues. Furthermore, the three groups were found to differ from each other more in terms of their perceptions of the negative consequences (costs) than the positive consequences (benefits). The findings presented in this section indicate that proponents and opponents have different views on the possible benefits and costs of a possible nuclear power station in their locality. Overall, a majority tends to oppose local construction and their views tend to be dominated by health and safety issues. Economic benefits seem to play a less important role, especially for those who oppose the building of a local nuclear power station.

16

Table 1.3 Perceived consequences of a proposed nuclear power
plant as a function of attitude

Potential Outcomes	Attitude		
	Pro (n=82) %	Neutral (n=43) %	Anti (n=45) %
Economic factors			
More electricity available	85	67	78
Slower increase in costs of electricity	44	49	33
More local jobs	61	65	69
More local business	43	53	36
Environmental factors and health risks			
Pollution (thermal, visual etc.)	72	74	96
Radiation hazard	83	56	91
Disturbances during Construction	18	19	38
Too much industrial development	19	14	40

Note: Scores represent the percentage of each group of subjects, who either voiced
or indicated an awareness of each issue.
Adapted from Woo and Castore (1980), p. 228.

The importance of health and safety issues is also apparent in
survey research. Unfortunately only a limited amount of survey data
is available that deals directly with questions of health and safety.
Nealy *et al.* (1983) report a number of surveys in which respondents
were asked to give the reason for holding a particular point of view.
Results showed that between 50 per cent and 75 per cent of those
who opposed nuclear power gave danger-related reasons for holding
such a view. Other surveys revealed similar findings. The harmful
consequences that are mentioned most often by respondents are
reactor accidents; routine low-level radiation, waste management and
the possibility of adverse health effects for present and future

generations (Nealey, *et al.* 1983, p.69). Canadian surveys reveal a similar concern about safety issues; 43 per cent of the population stated that they were simply unwilling to live with any risk of a nuclear accident (Decima, 1987). This survey also confirmed the earlier reported finding that spontaneous responses to nuclear power tend to be dominated by safety issues.

Summarizing, local acceptance seems a function of the (im)balance between health and safety hazards versus economic benefits. This brings us to the concept of equity, which is of particular importance in the local context.

Local attitudes and equity

As indicated in the previous section local acceptance of nuclear facilities is low and tends to be dominated by safety issues. Initially, attempts to explain these reactions focused on an 'irrational fear' on the part of the public. Some referred to 'panic' and 'irrationality' about radiation and chemicals (e.g. Cohen and Lee, 1979; DuPont, 1980; Bord *et al.*, 1985). Pahner (1976) suggested that public concern over nuclear power risks stems from images of the horrors of nuclear war, fears related to the invisibility of radiation, uncertainty about exposure, and the effects of radiation on genetic processes.

The presence of fear in public reactions towards nuclear waste is indisputable. However, relevant research does not suggest that fear plays a dominant role (Cunningham, 1985; Freudenburg and Baxter, 1984; Mitchell, 1984). Brown, Henderson and Fielding (1983) found concern (a cognitive concept) about nuclear power to be far more pervasive than anxiety (emotion). Johnson (1987) notes that despite the lack of evidence for irrational fear as the primary factor in attitudes towards hazards such as those of nuclear waste, one should not dismiss it entirely. Fearful aspects of personal or family exposure to radiation risks are obviously more salient for local residents. Dismissing public concerns as entirely irrational, however, is not likely to help our understanding of local reactions. Instead, more attention should be paid to concern about safety issues and risk-benefit equity as important elements of public opposition. The siting processes of nuclear power facilities have a major characteristic

18

which is possibly of even more importance than the safety aspect: inequity. Inequity, it is widely held, is a crucial problem for siting nuclear facilities. Indeed, for many it is *the* problem. Once a site is identified, the community tend to oppose the proposed facility because the *benefits* will flow to the owner, operator and the public at large while the *risks* will be concentrated locally. This seems especially the case for nuclear waste repositories which tend to create far fewer job opportunities than nuclear power stations. It has been argued for some time that any facility providing diffused benefits while imposing high costs at or near the specific location are bound to engender local opposition (see e.g. Kasperson, 1985). Generally, an acceptable balance of risks and benefits will be very difficult to obtain. From a local perspective, however, it seems fairly rational to oppose construction of facilities that introduce substantial risks for health and the environment and only limited economic benefits.

One could improve equity by increasing the economic benefits to compensate for the risks. There is some evidence suggesting that economic benefits can increase local acceptance. For instance, members of TMI-area citizen groups in favour of restarting the undamaged second reactor were generally in a position to benefit economically from the restart (e.g. shopkeepers); opponents were not (Soderstrom, Sorensen, Copenhaver and Carnes, 1984). Similarly, the *absence* of local benefits may increase local opposition. A study on local perceptions of the West Valley nuclear reprocessing plant in the US illustrates this point (see Kates and Braine, 1983). When the local population realized that the promised benefits of the facility were less certain, and other economic developments (for example tourism) might be harmed by the growing controversy, local concern and opposition to the proposed facility increased significantly.

Equity thus seems an important value underlying local attitudes and acceptability. Equity, however, is not the only value which is of importance in understanding attitudes to nuclear energy. In the next section we will briefly discuss the role of more general beliefs and values in the nuclear debate.

Attitudes, beliefs and values

Van der Pligt, Van der Linden and Ester (1982) approached a large sample of the Dutch population; about half of the sample lived close to an existing nuclear power station, the other half in comparable, rural areas *without* a nuclear power station. In their study respondents were also asked to indicate the importance they attached to a variety of more general issues. Table 1.4 presents these issues and the differences between pro- and antinuclear respondents.

Table 1.4   Importance of general values of pro- and antinuclear subjects (I).

		Pronuclear subjects (n = 179)	Antinuclear subjects (n = 349)
a)	Maintaining the present material standard of living	31	28
b)	Improvement of the strength of trade and industry	76	37
c)	Conservation of the natural environment	28	58
d)	Reduction of the level of unemployment	64	71
e)	A stricter criminal law	35	24
f)	Providing a less complex society	17	29
g)	Greater public participation in decision making.	03	12
h)	Increase of defence spending	24	03
i)	Reduction of energy use	16	26

Note: Scores represent the percentage of subjects selecting each issue among the three most important. The columns do not add up to 300 because of the inclusion of subjects who chose fewer than three issues.
Adapted from Van der Pligt, Van der Linden and Ester (1982), p. 226.

Their results showed that pronuclear respondents stressed the importance of economic development, whereas antinuclear

20

respondents put more emphasis on conservation of the natural environmental and the reduction of energy use. The antinuclear group also thought the issue of increased public participation in decision making to be more important (see Table 1.4).

A stepwise discriminant analysis was conducted on these scores to find the factors that *most* distinguished the pro group from the anti group. The results revealed three factors: 'improvement of the strength of trade and industry', 'increase of defence spending', and 'conservation of the natural environment'. Not surprisingly, a relation was found between these value differences and subjects' position on a political left–right dimension. Political preference was significantly related to the perception of these values, *and* to respondents' attitudes to the building of more nuclear power stations. Opinion poll surveys conducted when this study was being carried out confirmed this relationship between political preference and attitudes towards nuclear energy (see Van der Pligt *et al.*, 1982).

Table 1.5   Importance of general values for pro- and antinuclear subjects (II).

		Percentage of respondents selecting each aspect	
		Pro-subjects	Anti-subjects
a)	Decreased emphasis on materialistic values	36	100
b)	Higher material standard of living	36	0
c)	Reduction in scale of industrial, commercial, and governmental units	22	86
d)	Greater public participation in decision making	50	66
e)	Industrial modernization	68	6
f)	Advances in science and technology	82	13
g)	Security of employment	77	40
h)	Improved social welfare	31	80
i)	Conservation of the natural environment	77	100

Note: Adapted from Eiser and Van der Pligt (1979), p.532.

These findings from a general population sample resemble those of a smaller study (Eiser and Van der Pligt, 1979) in which we

21

compared two groups of relative 'experts' (nuclear industry employees and active environmentalists) attending a one-day workshop on 'The Great Nuclear Debate'. In this study subjects were asked to select the five aspects (out of nine) which in their opinion would contribute most to improving the overall quality of life. Table 1.5 summarizes the findings.

These results show clear differences between the two attitude groups, with pronuclear subjects stressing the importance of 'advances in science and technology', 'industrial modernization', 'security of employment' and 'conservation of the natural environment'. The antinuclear subjects put even more emphasis on the last aspect, and also stressed the importance of 'decreased emphasis on materialistic values', and 'improved social welfare'. These results indicate that attitudinal differences towards nuclear energy are embedded in a wider context of more general values and views on important social issues. Public thinking on nuclear energy is thus not simply a matter of perceptions of risks and benefits but is also related to more generic issues such as the value of economic growth, high technology and centralization (see also Kasperson *et al.*, 1980). It seems impossible, therefore, to detach the issue of nuclear energy from questions of the kind of society in which we want to live.

Conclusions

In this chapter we have presented a brief overview of public opinion research on the issue of nuclear energy. Results show that public opinion about nuclear energy remained relatively stable in the mid-1970s and become more antinuclear in the late 1970s. The 1979 accident at TMI accelerated this change and public opinion in the US became more antinuclear. Moreover, public support for nuclear power did not return to pre-TMI levels. The Chernobyl accident in 1986 had a similar impact on public opinion. This impact was most dramatic in a number of European countries and also resulted in more antinuclear opinions that remained relatively stable. Thus, it has been shown that significant nuclear accidents can influence attitudes towards nuclear power. It seems that the effects of such accidents dissipate more quickly when the spatial distance from the accidents is large and when the consequences are limited. For instance, return of public acceptance to pre-TMI levels was probably more complete in Europe than in the US (see for instance, Midden and Verplanken, 1986, 1990). On the other hand, whereas negative effects of Chernobyl on public attitudes persisted in Europe a year after the accident, attitudes in the US appeared to have returned to previous levels. Overall, public opposition and support for nuclear energy seems evenly balanced both in the US and in Europe.

This is not the case for attitudes held by local residents towards the siting and construction of a nuclear power station in their vicinity. Support for local plant construction decreased gradually and significantly through the 1970s, while opposition increased. Local attitudes seem now firmly against the siting of nuclear power plants both in Europe and the US. Most research shows a majority of people being opposed to the building of a nuclear power station in their locality. Local opinion is the central theme of this book. For that reason we have presented a brief overview of international research on local attitudes during the siting and construction stages of nuclear power stations. This research provides more information about the perceptions of local costs and benefits underlying public opinion. The NIMBY (Not-In-My-BackYard) phenomenon seems quite rational if one adopts a local perspective. Antinuclear subjects perceive the costs to be larger than the benefits, while pronuclear subjects are more optimistic about the (mainly economic) benefits.

Conflicts between opponents and proponents of the local construction of a nuclear power station quite often take the form of accusations about missing important aspects and/or possible consequences. What determines the importance of consequences is, of course, the central question in controversies of this kind. It seems that the pro- and antinuclear groups tend to see different aspects of the issues as important or *salient*, and tend to disagree not only over the likelihood of possible local consequences but also over their importance. Research described in this chapter shows that anti- and pronuclear respondents differ considerably in which consequences they see as most important. Economic and technical benefits made a more pronounced contribution to the attitudes of pronuclear subjects

23

while health, environmental, and socio-political risks were the prime determinants of the attitude held by antinuclear subjects.

Equity of costs and benefits is an important element of local opposition to nuclear power station construction. The lack of balance between local and national costs and benefits is an important factor underlying the limited acceptability of nuclear energy facilities. Equity is not the only relevant factor, however. Nuclear proponents and opponents also differ in their perception of more general values. This is the case for the general population as well as 'experts'. Antinuclear subjects appear far more committed to the philosophy of 'small is beautiful', and less convinced of the benefits of technological advances and a further improvement in material well-being. Questions of the kind of society in which one wants to live seem to preoccupy the antis far more than the pros, who seem to see the debate primarily as one concerning the adequacy of safety precautions. Attitudes towards nuclear energy are relatively stable and embedded in a wider context of values. Moreover, proposals for the construction of a nuclear power station will generally be opposed by local residents. Large scale nuclear accidents such as TMI and Chernobyl have affected both general and local attitudes but are not the main cause of local opposition. The imbalance between risks and benefits seems to be the prime determinant of local opposition. The findings presented in this chapter also point to the importance of the salience of the various possible consequences of the local building and operation of a nuclear power station. Anti- and pronuclear respondents adopt very different perspectives on the nuclear issue and attach differing importance to various possible consequences. We will return to this issue in Chapter 3. First, we will provide information about the overall design of the study conducted in the South-West of England and the context within which the research was carried out.

Chapter 2

Local Context and Research Design

Nuclear energy, like most environmental issues, has global implications. But it also has local implications and these, though often less newsworthy, may nonetheless be more significant for ordinary people living near to existing nuclear power stations or in areas where a new station is planned. If there is a plant already operating, what kind of contribution does it make to the local economy in terms of employment and the spending power of those employed? Is it regarded as an integral and valued part of the life of the local community or as a sinister and secretive intrusion? Is it seen as safe and well-managed or as a potential source of death and environmental disaster? If the concern is with the building of a new plant will this bring local economic benefits? Will the inevitable changes in terms of new roads, houses and services be seen as welcome or unwelcome developments? How will the character of the community change, both when the plant is operational and, more immediately, during the lengthy period of construction? Will it be an unsightly blot on the landscape or something one gets used to after a while? What would happen if anything went wrong? Why do *they* want to build it here?

These do not appear to be issues of national politics or macroeconomics in themselves, nor would most of those who ask such questions necessarily feel that they were engaged in any broader political or economic debate. Even so, the broader questions cannot be addressed if these more specific uncertainties are ignored. If there is a local community that *wants* its own nuclear power station (or an additional one), expansion of the nuclear industry becomes politically easier. If *no* local community wants such a development none can take place except by imposition by central government in the name of the 'national interest'. Thus community and national loyalties are put into opposition and this is something that is intensely political. Resistance at a local level can prove time-consuming and expensive for the industry in many ways. There are problems of public relations in handling the processes of consultation, of presentation of its case to a public inquiry. As the extent of local concern becomes apparent, plans or schedules may be revised to introduce extra safeguards and reduce still further the intrusiveness of any local environmental impact (which then will be demanded as minimum standards for any future developments elsewhere). Alongside all of this, there will be countless day-to-day management decisions depending on forecasts of when the new power station will be built, if at all, and when, if ever, it will start repaying its capital investment. These are economic issues of massive proportions.

Local attitudes, then, have far more than merely local significance. But if we are to understand them, we cannot ignore the historical and geographical context in which they are situated. In this chapter we shall describe the background to our own programme of research, conducted in the South-West of England at a time when proposals for a new nuclear power station in the region were under consideration. We shall then proceed to an overview of the design of the first stage of data collection, and report preliminary findings relating to the characteristics and attitudes of our sample.

A new power station for South-West England?

In February 1980, the Central Electricity Generating Board (CEGB)—the public body which at that time controlled all electricity generation in England and Wales—announced its intention to investigate five sites in South-West England as possible sites for a new nuclear power station. From an engineering point of view, the case for more generating capacity in the region appeared reasonable. The South-West peninsula is—in English terms—a relatively large geographical area with few large centres of population: Plymouth has roughly a quarter of a million inhabitants, whereas Exeter and Torbay each approach 100,000. However, there is a large influx of visitors during the summer months. The peninsula uses more electricity than is generated in the region and relies for its supplies on lengthy transmission lines. The implied insecurity of local supplies, so the argument first went, would be eased or removed by a new power

26

station. The long coastline and sparse population should also make it easier to find sites that would be easier to evacuate in an emergency—not that the actual need for an emergency evacuation was seen as more than a remotely hypothetical consideration.

The South-West already had a major nuclear facility at Hinkley Point on the North Somerset coast of the Bristol Channel. This consists of two power stations, one using the early generation of Magnox reactors, and the other an Advanced Gas-Cooled Reactor (AGR) of British design. However, by 1980 AGRs were already being viewed as commercially and operationally disappointing. The CEGB wanted a new generation of stations using American-designed Pressurized Water Reactors (PWRs). Plans had been announced (but not approved) for a new PWR at Sizewell on the Suffolk coast of East Anglia and it was anticipated that a PWR for the South-West would be next in line. Apart from Hinkley Point, generating capacity in the region was negligible, with older coal- and oil-fired stations (near Barnstaple, Plymouth and Poole) about to be taken out of operation. A significant exception was Winfrith in the county of Dorset, where the United Kingdom Atomic Energy Authority (UKAEA) has a research establishment, contributing power to the National Grid from a small reactor.

The CEGB's announcement of February 1980 was accompanied by the map shown in Figure 2.1. Two possible sites were identified on the North Cornwall coast, fairly close to Land's End. The most westerly, Gwithian, would certainly have been visible across the bay from the popular and exceptionally picturesque holiday resort of St Ives. Nancekuke is a little further to the east, around a headland from St Ives Bay. The site in question, skirted by the coastal footpath, belongs to the Ministry of Defence and is more easily located as RAF Portreath, a radar tracking station. Moving further east, we come to rougher rural countryside a little inland between the villages of Luxulyan and Bugle. The landscape immediately to the west of Bugle resembles a white-washed version of the South Wales coalfield, for this is a main centre of the Cornish china clay industry, and the waste tips rise as incongruously as a lunar landscape (for which indeed they are said to have doubled in the occasional science fiction film). Turn east towards Luxulyan, though, and rolling farmland reasserts its sovereignty.

27

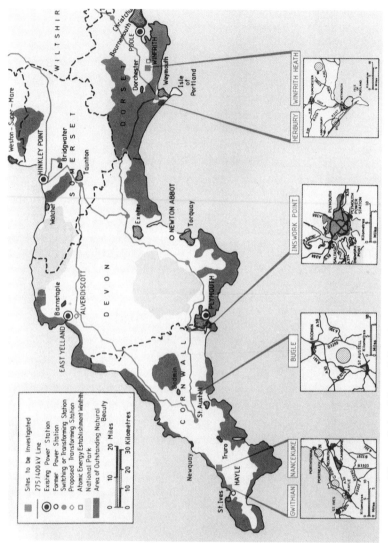

Figure 2.1   Map of South-West England showing the proposed sites

Right up against the city of Plymouth, an arrow identifies a possible site at Inswork Point, but this was not being considered for a nuclear reactor. This site had been originally acquired by the CEGB for a 1320 MW oil-fired station, for which planning permission was obtained in 1973. However, five years later, investment approval was withheld by the Labour government on the grounds of national energy policy and by 1980 the increase in world oil prices had undermined the viability of the original plan. This site was now to be reassessed in terms of its suitability for a coal-fired power station instead. Whether or not this was regarded as a serious alternative to nuclear expansion, little has since been heard of this proposal.

Passing east along the south coast, past Devon and into Dorset, we come to a site known as Herbury. On most other maps, this is hard to find, but the site in question lies near Langton Herring, a well-groomed village of tidy stone cottages. Just below the village is an eight-mile (13 km) long spur of sand and shingle known as the Chesil beach, which shelters a strip of water between land and sea known as the Fleet. Part of an officially designated 'Area of Outstanding Natural Beauty', this is rich in bird life, with a swannery at Abbottsbury at its northern end. Herbury Gore, from which the proposed site took its name, is a grassy hillock on the Fleet's inner shore. The final site is of rather less dramatic beauty but still has a special attractiveness and importance. Winfrith Heath is an impressive area of open uncultivated heathland. It is rich in flora and fauna, and for this reason much of it is designated as a 'Site of Special Scientific Interest'. The heath adjoins the research establishment at UKAEA Winfrith, and so some case could be made that this last proposal would merely involve the expansion of an existing facility. The fact that this existing facility is located in a rare area of lowland wilderness makes the prospect of any such expansion less trivial ecologically.

None of these sites, then, would be without environmental costs, but the CEGB stressed that the choice of site would be subject to consultation with local residents and other interested parties, and would in any event require investigations for geological suitability. On the other hand, it was clearly the CEGB's committed view that a new station was needed in the region, and quite probably more than one in the medium term. There was considerable doubt about where, some doubt about when, but almost no doubt about whether the South-West would get its PWR.

The choice of site

Consultations and geological tests then proceeded, though not without some difficulties. Villagers at Luxulyan achieved possibly the most publicity with a sit-in protest in a field designated for test drilling, to be removed by the Devon and Cornwall Constabulary only after a court order and with a transparent lack of enthusiasm. However, there was no single pattern to the actions of local objectors and—depending on how one interprets the events at Luxulyan—no significant acts of civil disobedience. Predominantly, the issue seemed to be defined in local terms, and few residents of these communities would have regarded themselves as the front line of some broader anti-nuclear movement.

Two years after publication of the initial proposals, the CEGB made an interim announcement in February 1982 that, because of 'unsatisfactory geological conditions', it was excluding the two West Cornwall sites (Gwithian and Nancekuke) from further consideration for the foreseeable future. The spotlight then turned onto Luxulyan and the two Dorset sites, but when the announcement of the final selection came on August 25th 1982, a dramatic change of direction was revealed. The CEGB's press release presented the decision both as part of a national strategy and as sensitive to local interests:

> The Central Electricity Generating Board has now completed its assessment of potential power stations in South West England and these and other sites have been considered for future development...
>
> After more than two years of investigations, environmental studies and consultations, the work has now been completed and the Board is able to announce the power station sites it considers most suitable for development in the South West and elsewhere in England...
>
> Against this background the next station for which planning application will be made will be on land adjoining the existing power stations at Hinkley Point in north Somerset. It will be known as Hinkley Point 'C'...

> The site considered at Winfrith Heath would be technically suitable for power station development. However, the Board considers that an area largely within the site of the UKAEA establishment itself would be more suitable for power station construction purposes taking into account ecological aspects... The Board will not proceed further with the Winfrith Heath site originally under consideration...
>
> The locations investigated at Luxulyan... and at Herbury... have also been proved to be technically suitable for power station construction. At Luxulyan, whilst a new power station could be set to some extent against the background of the extensive clay workings, it would still be dominant over a large area, and a station development would be much more costly than at the alternative sites... At Herbury... the site itself and its surroundings are of intrinsic natural beauty and ecological value which strongly weigh against its selection for development. In view of these considerations, provided satisfactory progress can be made with planning new stations at the Hinkley Point and Winfrith AEA sites, the Board does not intend to proceed further with either Luxulyan or Herbury.

The press release explained that the sequence of events would be to await the outcome of the public inquiry into the plans for a PWR at Sizewell B. Planning permission would then be sought for Hinkley Point C (either for an AGR or a PWR, but it was an open secret that it would be a PWR unless something unforeseen happened at the Sizewell inquiry). Thereafter 'proposals for the next consent application' would be chosen from four named sites, including Winfrith (that is, largely on land already within the area of the UKAEA research establishment). The other three sites were elsewhere in the country: at Druridge, Northumberland (a new site) and for additional stations to those already planned at Sizewell and at Dungeness, Kent.

With regard to non-nuclear facilities, the Inswork Point site (and a similar one at Marchwood near Southampton) were to be 'retained to provide options for longer term development'. The existing sites at Barnstaple, Plymouth and Poole would not be redeveloped.

Thus, the strategy was to make maximal use of existing sites and to avoid controversial developments where no previous power stations had been built (although Druridge was a notable exception). On the other hand, the press release refers to new *stations* in the

31

plural. Winfrith is still clearly part of a longer term strategy, and even the reprieve of Luxulyan and Herbury carries the proviso 'provided satisfactory progress can be made at the Hinkley Point and Winfrith AEA sites'. In short, there is a strong commitment to a strategy of expansion and the only kind of development seriously considered is nuclear.

Over the following several years, the CEGB developed detailed plans for Hinkley Point C, which were submitted to a public inquiry which started in October 1988. In the meantime, the CEGB had gained approval for the building of Britain's first PWR at Sizewell B. The CEGB had viewed the Sizewell public inquiry as one that would make a 'generic' decision regarding the economic, operational and construction aspects of the PWR programme as a whole, so that subsequent inquiries, such as that concerning Hinkley Point, need only concern themselves with issues of specifically local impact. We shall consider some aspects of the Hinkley Point inquiry in Chapter 7. As for 'satisfactory progress' at Winfrith, technical problems have since forced the UKAEA to announce the early closure of its plant. The likelihood of development of the site to include a new CEGB reactor may therefore be regarded as remote.

Overview of the research design

The research to be described in this and subsequent chapters was supported by a grant from the Economic and Social Research Council, London. Because of this, we were able to approach participants as working independently of any political, governmental or commercial institution. This was an undoubted advantage. Most of the data we obtained came from postal questionnaires, sent to random samples from the relevant communities. Significantly, respondents were asked to record their names. We explained that this was so that we would know that they had responded and would not need to send them a reminder. Another vital advantage for us was that we could build longitudinal comparisons over time into our design. Our assurances of confidentiality were clearly trusted, judging from the high response rates (to be described). Doubtless, in some kinds of research anonymous responding may be essential to the validity of the data. However, we suspect that some surveys which lay unnecessary emphasis on anonymity may convey the impression that genuine viewpoints can only be expressed secretively.

The questionnaires we distributed were composed of several sections, addressing sometimes rather separate research questions, and with the content of some sections systematically varied to enable particular experimental comparisons to be made. For this reason, subsequent chapters will be organized around specific research questions, rather than around single samples or phases of data collection, except where these happen to coincide. In this chapter, however, we shall retain a more orthodox structure, describing our original design and simpler group comparisons. Following this, there will be a summary of how our later studies related to our first.

The 'four counties' study

Timing

The original conception of our research was to compare attitudes and beliefs among residents of communities near the five sites named in the 1980 announcement, as well as of two control communities, up to and after the final choice of site. Our project started at the beginning of January, but not before it was rumoured that an interim announcement to reduce the shortlist was imminent. We therefore adjusted our plans to allow for data collection at three points in time, specifically:

Time 1 (January 1982)—before the interim announcement;
Time 2 (March to April 1982)—after the interim announcement;
Time 3 (September to October 1982)—after the final announcement.

Irrespective of locality, three equivalent random samples were drawn. Group *A* was to be contacted at time 1 and recontacted both at time 2 and time 3, with earlier and later data from the same individuals matched by name. Group *B* was to be contacted for the first time at time 2 and then again at time 3. Group *C* was to be contacted for the first and only time at time 3. Thus comparisons could be made across

33

time in a manner that controlled for the effects of repeated testing. This design generated the following data sets:

1-A at time 1;
2-A and *2-B* at time 2;
3-A, 3-B and *3-C* at time 3.

Locality

The effects of locality were compared by sampling from the electoral registers of small towns within approximately a 15 to 30 km radius of the relevant sites. Because of the proximity of the two sites in West Cornwall and the two in Dorset, this resulted in three 'districts' facing the prospect of a new power station, to which two control 'districts' were added. Each of these five districts included a mix of agricultural, market town and seaside neighbourhoods, with a fair proportion of retired people. The five sampling districts so identified were:

West Cornwall, covering the Gwithian and Nancekuke sites and drawn from the towns of St Ives, Hayle, Camborne and Redruth.
East Cornwall, covering the Bugle/Luxulyan site and drawn from Bodmin, St Austell and Fowey.
Dorset, covering the Herbury and Winfrith sites and drawn from Dorchester and Weymouth.
Somerset, drawn from Bridgwater and Highbridge. This was chosen to be a control sample, not involved in the 1980 announcement, but already familiar with a nuclear power station (Hinkley Point) nearby.
South Devon, drawn from Newton Abbott, Dawlish and Teignmouth. This was chosen as a control sample, not involved in the 1980 announcement and without an existing nuclear plant nearby.

As should be apparent, we failed to anticipate the transformation of our 'control' group from Somerset into the 'preferred site' group as a consequence of the final announcement. Even so, at times 1 and 2, the Somerset residents may be regarded as a comparison group along the lines originally intended, in that they were familiar with a nuclear plant in their locality, but had not been led to expect further development of this facility.

	Feb 80	Jan 82	Feb 82	Mar–June 82	Aug 82	Sept–Oct 82
CEGB initial announcements			interim		final	
Study phases		*Time 1*		*Time 2*		*Time 3*
District samples						
West Cornwall	}	*1 - A* ---------------->		*2 - A* ---------------->		*3 - A*
East Cornwall	}					
Dorset	}			*2 - B* ---------------->		*3 - B*
South Devon	}					
Somerset	}					*3 - C*
Village samples						
Luxulyan	}			*2 - V* ---------------->		*3 - V*
Langton Herring	}					
Winfrith	}					*3 - W*
Hinkley Point	}					*3 - X*

Figure 2.2 Overall design of the 'four counties' study

An important modification to this design was decided upon when planning the data collection at time 2. From visiting the communities concerned, it occurred to us that we might have been sampling over too wide a geographical area to pick up some of the special concerns possibly felt by residents of the villages immediately adjoining the proposed sites. We therefore expanded our design to include residents of the three villages most immediately under 'threat' following the interim announcement. These were Luxulyan in Cornwall and Langton Herring in Dorset (already mentioned) and that of Winfrith Newburgh in Dorset, close to UKAEA Winfrith. No 'village' samples were included from the West Cornwall sites, which by then had been excluded by the CEGB's February 1982 announcement. Once again, we controlled for possible effects of repeated testing by selecting one sample, group V, to be tested both at time 2 and time 3 (generating data sets 2-*V* and 3-*V*), and another, group *W*, to be tested only at time 3 (generating data set 3-*W*). The CEGB's final announcement, however, left us without a 'village' sample close to the site that was actually chosen. We therefore added a final sample (3-*X*), drawn from residents of small hamlets between Bridgwater and Hinkley Point. The overall design is summarized in Figure 2.2.

Response rates
The sampling procedure was based on the electoral registers of the relevant communities, the final samples being drawn by first selecting electoral wards at random (from the district but not village communities, since the latter were typically contained within one or two wards) and then selecting names at random from within the wards. Questionnaires were then dispatched by 'recorded delivery', meaning that they were only delivered if the named individuals were at the address specified.

Table 2.1   Response rates: 'four counties' study

Data set	Sent	'Gone away'	Returned		Rate (per cent)
1-*A*	750	67	356	(7)[a]	52
2-*A*	349	10	257	(9)	76
3-*A*	248	9	192	(0)	80
2-*B*	491	67	250	(14)	59
3-*B*	236	5	148	(2)	64
3-*C*	375	42	202	(30)	64
2-*V*	450	24	300	(10)	70
3-*V*	290	2	189	(6)	66
3-*W*	300	20	187	(11)	67
3-*X*	400	25	232	(14)	62

[a] Figures in brackets refer to unidentified returns

We therefore were able to distinguish between non-respondents and cases where the individuals had moved home or had died since the registers were compiled a year before. Such cases are referred to as 'gone away'. As mentioned, respondents were asked to record their names and addresses on the front page of the questionnaire. This enabled us to contact all non-respondents to any of the first mailings, typically after a lapse of about two or three weeks, with a reminder letter and a second copy of the questionnaire. All mailings provided a 'freepost' reply envelope on which no stamp was required. The response rates yielded by this procedure are shown in Table 2.1. As may be seen, they are all gratifyingly high.

Characteristics of the sample

All respondents were asked to provide some basic demographic information, including sex, age, their intention to remain in the area and their perceived job mobility, whether they owned their own home, and their present or previous occupation and/or that of the main 'breadwinner' in their family. Reported occupation was coded, according to the Registrar General's system of classification, into five categories of socio-economic status from 1 (professional) to 5 (unskilled). These questions were included in the first questionnaire each respondent personally received (i.e. 1-A, 2-B, 3-C, 2-V, 3-W, 3-X). However, 460 members of the A, B and V groups were not retained at time 3, having failed to provide identifiable follow-up data. This left 1087 respondents, of whom 54.9 per cent were male. Their average age was 47.0 years (SD = 18.4) and 65.3 per cent owned their own home. Classification of socio-economic status yielded the following percentages for the respective categories: 1, 7.9; 2, 17.2; 3, 42.8; 4, 22.4; 5, 9.7 (139 respondents could not be classified).

Perceived job mobility was assessed by the question 'Could you (or the main "breadwinner" in your family) get a similar job elsewhere if you moved away from the area?' The distribution of responses in terms of five categories were 'yes, definitely' 23.1 per cent, 'yes, probably' 34.1 per cent, 'don't know' 13.4 per cent, 'probably not' 15.6 per cent, 'definitely not' 13.7 per cent. These responses were significantly associated with locality , combining 'village' samples with their corresponding 'districts' ($p < 0.02$). Rather low job mobility was reported by East Cornwall residents, with 44.7 per cent saying that they could definitely or probably move their job, but 38.6 per cent saying that they definitely or probably could not. In Somerset, the corresponding figures were 64.1 per cent and 24.6 per cent, and in Devon, 67.8 per cent and 16.1 per cent.

A related question—albeit more attitudinal than strictly factual—asked 'Would (or does) having a nuclear power station nearby make you want to move away from the area?' Responses were in terms of the same five categories, and distributed as follows: 'yes, definitely' 13.6 per cent, 'yes, probably' 12.8 per cent, 'don't know' 8.9 per cent, 'probably not' 28.6 per cent, 'definitely not' 36.1 per cent. Again, these responses were significantly associated with locality ($p < 0.001$). Among East Cornwall residents, 46.7 per cent said they would definitely or probably want to move, with 43.9 per cent saying they would not. In Somerset, the corresponding figures were 9.4 per cent and 76.2 per cent.

Expectations concerning the choice of site

A question which tracked the effects of the CEGB announcements was one which asked for estimates of the likelihood of a new nuclear power station being built at any of the five original sites 'in the foreseeable future'. Hinkley Point was added at time 3 to the list of sites presented. Ratings were in terms of a seven-point scale from 'almost no chance at all' (1) to 'almost certain' (7). This question was asked of samples *A* and *V* only (data sets 1-*A*, 2-*A*, 3-*A*, 2-*V*, 3-*V*). Table 2.2 shows the mean likelihood assigned to each site at each time for samples *A* and *V* as a whole, that is, without regard to respondents' own place of residence. As may be seen, Winfrith was favourite on both the first two occasions and was still regarded as quite likely after the final announcement (which held open the possibility of a further power station being built at Winfrith after that at Hinkley Point).

As one would expect, the judged likelihood of the two sites in West Cornwall was noticeably less after the interim announcement (between times 1 and 2) that officially excluded them, although the

38

means (still higher than the minimum of 1.0) might imply some lingering doubt over the finality of the 'reprieve'.

Table 2.2 Perceived likelihood of a new power station at each site over time

Site	District samples			Village samples	
	Time 1	Time 2	Time 3	Time 2	Time 3
Gwithian	3.3	1.9	1.7	1.9	1.7
Nancekuke	3.6	2.0	1.8	2.0	1.8
Bugle/Luxulyan	3.6	4.0	2.3	4.1	2.6
Herbury	3.5	4.2	2.6	4.3	2.5
Winfrith	4.3	4.6	4.7	4.8	5.4
Hinkley Point			6.0		6.1

Note: Time 1 = January 1982; Time 2 = March–June 1982; Time 3 = September–October 1982; scores range from 1 ('almost no chance at all') to 7 ('almost certain').

Similarly, the ratings for the Luxulyan and Herbury sites, though they drop from time 2 to time 3, remained well above the minimum even after they were finally excluded. This reluctance to revise prior estimates so as fully to reflect new evidence may be an example of what Tversky and Kahneman (1974) have called the 'anchoring and adjustment' heuristic: estimates remain 'anchored' to prior judgments based on initial evidence and tend to be only partially revised when new evidence becomes available. However, there may have been another contributory explanation. The 'evidence' on which these estimates were revised consisted of announcements by the CEGB who, as an interested party, might have been seen as a less than completely trustworthy source of information by some respondents (see Chapter 5). Furthermore, the final announcement mentioned the technical suitability of the Luxulyan and Herbury sites, and hinted at a renewal of interest in these sites in the absence of 'satisfactory progress' at Hinkley Point and Winfrith.

General attitudes towards nuclear energy

Although there were several different versions of the questionnaire, all contained a section containing eight attitude statements designed to measure attitudes to nuclear energy in general. Responses were in terms of a seven-point scale from 'very strongly disagree' to 'very strongly agree'. The statements read:

1. Nuclear energy is the only practical source of energy for the future.
2. Nuclear power stations are far cleaner than any other kind of power station.
3. Alternative technologies such as solar, wind or wave power are a far more sensible investment than nuclear power.
4. Nuclear energy is vital to the country's economic future.
5. Even scientists have little idea what would happen if there were a major accident at a nuclear power station.
6. Overall, science and technology have greatly increased the quality of life for ordinary people.
7. Nuclear energy is far less important than its supporters claim.
8. Britain should abandon all plans to build any more nuclear power stations.

Item analysis indicated that reliability would be improved (to around 0.85) by exclusion of item 6 from any combined measure. Accordingly, the remaining seven items were combined to form a single (Likert) attitude measure, hereafter referred to as the *attitude index*. Items 1, 2 and 4 were scored from 1 to 7 so that higher scores represented greater agreement, items 3, 5, 7 and 8 in the reverse direction, yielding an overall attitude index from 7 (extremely antinuclear) to 49 (extremely pronuclear). (In fact, in some analyses on data sets described in subsequent chapters, item 6 proved less problematic, allowing the use of an index based on all eight items, with a range from 8 to 56.) Taking the sample as a whole, there were no differences in general attitude as a function of socio-economic status. However, women tended to be more antinuclear than men, a finding consistent with most survey research (e.g. Nealey *et al.*, 1983).

Attitudes over time

One of the first things we looked at was the possibility of a shift over time in general attitude. Certainly it would have been the hope of the CEGB that consultations, public meetings and other forms of presentation of more detailed information would have allayed many of the misgivings of local residents, producing a more favourable shift over time. However, this obviously rests on the questionable assumption that such publicity would have led to greater acceptance, rather than increased resistance, as more residents became involved in their opposition to the proposals. What we could not assess, unfortunately, was the extent of any attitude change that might have taken place over the two years from the time of the CEGB's original announcement until we were able to commence our research. In any event, this aspect of our longitudinal design revealed essentially no differences as a function of time, nor were there any differences between the attitudes of individuals who responded on more than one occasion and those who failed to respond to later questionnaires. The sample means ranged from 22.5 to 27.7. Thus all groups were on average somewhat to the unfavourable side of the nominal scale midpoint of 28, although it is always dangerous to interpret such midpoints as representing absolute neutrality.

Differences in attitude between localities

As far as locality differences in overall attitude are concerned, the data collected at time 3 provide the fullest comparisons. Combining the data from all time 3 samples (3-*A*, 3-*B*, 3-*C*, 3-*V*, 3-*W* and 3-*X*), that is putting the village samples together with their respective districts, we find the following means on the seven-item attitude index, which differed highly significantly ($p < 0.001$) overall: West Cornwall 23.3, East Cornwall 26.6, Dorset 26.9, South Devon 26.9, Somerset 27.5. Combining these groups, there were no differences in general attitude as a function of socio-economic status or of age. However, men tended to be more pronuclear than women (means: 26.0 vs. 24.7, $p < 0.02$).

41

Another way of considering these data is by splitting the sample at the median into a relatively 'anti' group with scores of twenty-five or below, and a relatively 'pro' group with scores of twenty-six or above. The resulting frequencies are shown in Table 2.3. As indicated by the means above, there were significant differences between localities in the combined frequencies for district and village samples ($p < 0.001$), with Cornwall being more antinuclear than the remaining counties ($p < 0.001$). Although 'village' residents, considered as a whole, were no more antinuclear than 'district' residents, this disguises important differences between localities. In East Cornwall ($p < 0.05$) and Dorset ($p < 0.001$), there were proportionately more opponents of nuclear energy among the 'village' residents. Villagers from Langton Herring were slightly more antinuclear than those from around Winfrith, but this was not statistically reliable.

There is some evidence, therefore, that proximity of residence to a full-scale nuclear power station already in operation was associated with more favourable attitudes to nuclear energy in general. A possible interpretation of this finding is that the actual experience of a local station was experienced as less disturbing (and/or more directly profitable in terms of business or employment) than that feared by residents from other localities. This would be consistent with the finding reported earlier, that Somerset residents reported less actual wish to move away than those from other localities said they would feel, *if* there was to be a new power station in their neighbourhood.

Table 2.3   Attitude to nuclear energy as a function of locality: frequencies below and above median.

Districts	Anti	Pro	Villages	Anti	Pro
West Cornwall	53	37	Luxulyan	76	24
East Cornwall	64	42	Langlon Herring	66	62
Dorset	27	68	Winfrith	49	65
South Devon	46	59	Hinkley Point	71	130
Somerset	41	54			

An alternative interpretation, however, could relate to self-selection as a contributory factor. People who were antinuclear in inclination might be less likely to choose to live near a nuclear power station and might be more likely to move away, if they had the opportunity. Some support for this comes from responses to the question previously discussed concerning wish to move from the area. Although only 26.7 per cent of those responding indicated that they would probably or definitely want to move away if a new power station was built, the wish to do so was strongly associated with a more antinuclear overall attitude ($p < 0.001$). Other locality differences will be discussed in more detail in the next chapter (see section on 'Attitudes, familiarity and salience').

Guide to analyses reported in subsequent chapters

The findings just described are only a part of those obtained from the 'four counties' study, but are the main ones of interest relating to the sample as a whole. Our emphasis thereafter turned to looking, *within* the various groups, for variables associated with individual differences in attitude towards nuclear energy in general and a new local station in particular.

As mentioned in Chapter 1, there are many examples in previous research of the so-called 'NIMBY'(Not-In-My-Backyard) effect: opposition tends to be more intense towards nuclear and other developments in one's own neighbourhood. In Chapter 3 we will look at these local reactions in more detail and will focus on how specific risks and benefits are evaluated by local residents. The simple prediction—in keeping with much conventional theory in the fields of attitude research and of economic decision-making—is that residents who are more opposed to having a new power station in their neighbourhood think it would be more disadvantageous and less beneficial overall. Behind this overall relationship, though, one can look for more interesting associations between beliefs about particular kinds of consequence on the one hand, and particular degrees and styles of acceptance or rejection of the CEGB's plans on the other. Based on findings obtained in sample 2-V (see Figure 2.2), we shall compare attitudes towards the building of a local power station with

43

attitudes towards nuclear energy as such. We shall also look at the perceived importance, or 'salience' of particular anticipated consequences for different attitude groups. In the latter part of Chapter 3, we shall describe analyses on the three 'village' samples concerned with the effects of 'familiarity'. We shall draw the conclusion that the prospect of a new nuclear power station is interpreted as a different kind of issue by residents with different attitudes and experience. Indeed, differences in attitude may in large part be attributable to such differences in interpretation.

Chapter 4 reports data from the 'four counties' study concerned with comparisons between nuclear and other sources of energy. These comparisons involved not only preferences, but also estimates of the present and future contribution of different sources of energy to the national electricity supply. The logic behind these questions was an expectation that acceptance of nuclear energy would be related, in part, to a belief that alternative sources—whether 'renewable' or fossil fuels—are less practicable and/or desirable. This chapter also describes data from later studies (conducted in 1985/6) comparing attitudes to nuclear development and to oil drilling on the Dorset coast. Parallels are drawn between the factors leading to opposition to a possible new local nuclear power station (Winfrith) and to local oil drilling. The importance of relative proximity of residence to either development is also investigated.

Chapter 5 is concerned with how the different sides in the nuclear debate characterize each other, as well as how nuclear and alternative energy sources are portrayed in the media. A content analysis is presented of the treatment of energy issues by local newspapers in different parts of the country during the first six months of 1981. We also deal here with how residents sought to explain the CEGB's choice of site: what assumptions did residents make about the factors that weighed most heavily with the industry, and of the susceptibility of the industry to different sources of influence, including government and the media. Another question addressed is how much people expect others to share or not share their own opinions, and how this may depend on possible biases and expectations.

Chapter 6 presents data concerning attitudes about nuclear accidents and hazardous pollution. This theme was not prominent in the questions included in the 'four counties' study, but was tackled more directly later. One set of data was obtained in late November 1983, shortly after a widely publicized television documentary concerning radioactive pollution around the nuclear fuel reprocessing plant at Sellafield in Cumbria (North-West England). The sample contacted for this study was all those who had responded thirteen to fourteen months earlier to the third phase of the 'four counties' study. Reactions to the Chernobyl accident (April 1986) were also assessed within two very different samples. The first was a group of Dorset residents who had been contacted just before the accident concerning their attitudes to nuclear and oil developments, and the second consisted of samples of university students from five countries.

Finally, Chapter 7 draws together implications of our research for an understanding of public attitudes to nuclear energy and technological risk in general. The issue of 'risk perception' will be discussed within the context of more general attitudes, involving judgments of the trustworthiness and competence of those on whom technological safety depends. Related to this is the question of how much weight should be given to public and/or expert opinion when deciding whether the level of risk associated with a given hazard may be regarded as acceptable, or at least as 'tolerable'. We shall criticize the approach (attempted, for instance, by the CEGB in its presentation of its case to the public inquiry into Hinkley Point C) of defining risk levels associated with a new power station in simple numerical terms, and deciding that such levels must be 'tolerable' if they are less than those calculated for other, presumably tolerated, activities or industries. Such an approach may tell us what some industries think members of the public ought to tolerate, but not what they actually will. To understand this latter question, we must look at processes of attitude formation and change, at the selective and sometimes idiosyncratic ways in which people weigh information, at how they estimate the benefits and costs of any development in terms of their own system of values, and at how they construe the relationships between their own community and 'the powers that be'. These are issues of attitude and of social psychological processes, and it is from the perspective of such processes that our findings will be described.

Chapter 3

Local Attitudes: Costs, Benefits
and the Salience of Consequences

In Chapter 1 we argued that local opinion about the construction of a nuclear power station in the area tends to be more opposed and more extreme than opinions about nuclear energy in general. We briefly discussed this NIMBY (Not-In-My-BackYard) phenomenon, and argued that the best way to study these local reactions is to relate them to perceived benefits and risks of the local construction and operation of a nuclear power station.

In this chapter we will study these local reactions in more detail. First, we will introduce the central concepts of this chapter: attitudes and salience. Next, we will compare local opinion to the building of a nuclear power station with attitudes to nuclear energy as such. The analyses will also deal with the salience of the various possible consequences of a local nuclear power station for different attitude groups. In the last part of this chapter we investigate the effects of 'familiarity', or experience with living near an existing nuclear power station, on the perception of local risks and benefits of the building of a local nuclear power station.

Attitudes and salience

The concept of attitude has a long history in social psychology and, as noted by Dawes and Smith (1985), has known many definitions. There is little agreement about the *precise* definition of attitude. Allport (1935) noted that 'attitudes are measured more successfully than they are defined' (p. 9). The 'three-component' view of attitudes, proposed by Rosenberg and Hovland (1960) has traditionally provided a general framework for the study of attitudes. In this approach an attitude is defined as having three components;

47

affect (concerned with feelings, evaluations and emotions), *cognition* (concerned with beliefs about whether something is true or false) and *behaviour* (concerned with behavioural intentions). Attitudinal research on the nuclear issue has generally paid little attention to the affective and behavioural component of attitudes and tends to focus on the cognitive elements.

The models most often used to study the structure of attitudes towards nuclear energy are all related to cognitive, expectancy-value approaches of the attitude concept. These approaches are based on Edwards' work on *subjective expected utility* (Edwards, 1954). Anticipated costs and benefits are the two major components of subjective expected utility (SEU) theory. The work of Fishbein and his colleagues (Fishbein and Ajzen, 1975; Ajzen and Fishbein, 1980) is an application of SEU-theory and has been very influential in the attitude literature of the last two decades. The fundamental proposition of Fishbein and Ajzen's expectancy value model states simply that 'a person's attitude towards any object is a function of his or her beliefs about the object and the implicit evaluative responses associated with those beliefs' (Fishbein and Ajzen, 1975, p. 29). Alternatively stated, the attitude towards an object is a function of the subjective probability that the object is related to a specific attribute multiplied by the evaluation of that attribute, summed over the relevant attributes.

Quite often these beliefs or attributes will be anticipated outcomes that could occur in relation to the attitude object. As argued before, there is a close relationship between the notions of the subjective expected utility concept, anticipated cost and benefits, and Fishbein and Ajzen's model. The sum of the subjective probabilities weighted by their evaluation can be viewed as the overall utility of an attitude object. Benefits are the summed products of positive attributes, costs the summed products of the negative attributes.

The assumption of the model is that attitudes are the result of subjective but 'rational' information processing; i.e., beliefs or attributes are formulated in terms of possible positive and negative outcomes and integrated into an overall evaluation or attitude. In other words, each individual is presumed to engage in a (simple) form of cost-benefit analysis. Although the various components are subjective, relatively crude, and non-quantitative, the process of coming to an overall attitude is assumed to be rational.

As mentioned earlier, this expectancy-value approach has frequently been applied to the issue of nuclear energy. Since the mid-1970s there have been quite a few studies attempting to analyse the structure of people's attitudes towards nuclear power. Two early examples are the studies by Otway and Fishbein (1976) and Otway, Maurer, and Thomas (1978). The most important conclusion of Otway and Fishbein (1976) was that an analysis of the cognitive structure underlying attitudes towards nuclear power could identify factors differentiating between people with favourable and unfavourable attitudes. The major difference between pronuclear and antinuclear subjects was related to differing beliefs about the possible benefits of nuclear energy. Consistent with this finding, the benefit-related attributes were most important for the pronuclear group while safety-related aspects were most important for the antinuclear group.

Otway and Fishbein (1976) argued that much of the conflict surrounding nuclear power is due to different perceptions and that expectancy-value approaches provide a relatively objective measurement and description of these perceptions through the identification of beliefs or attributes used by different groups. In a later study, Otway, Maurer and Thomas (1978) report the results of a factor analysis on 39 attributes of nuclear power. For the pronuclear subjects, the factor of economic and technical benefits made the most important contribution to their overall attitude; for the anti group, the risk factors were more important. The anti group believed that environmental and health risks would be increased by the use of nuclear power, while the pro group believed they would not.

In Chapter 1 and in this chapter we summarized a number of studies showing that separate dimensions of the nuclear energy issue appear differentially salient for the different attitude groups. These studies focused on local attitudes during the siting and construction of a nuclear power station. As we will see later, the finding that separate dimensions of the issue of nuclear energy are differentially salient (in terms of their contributions to the prediction of overall attitude) for the different attitude groups, could have important practical implications for communication in the nuclear debate.

In this chapter we shall briefly describe a number of studies showing that separate dimensions of the nuclear energy issue appear differentially salient for the different attitude groups. This chapter is largely based on studies reported in Van der Pligt, Eiser and Spears (1986a, 1986b) and Eiser, Van der Pligt and Spears (1988). In the next section, we will describe our research on this issue conducted in South-West England.

Salience and local attitudes

In this study we investigated the perception of a variety of possible consequences of the building of a nuclear power station in South-West England by different attitude groups. The attributes or consequences included a wide variety of short-term and long-term consequences of the building and operation of a nuclear power station in the region (see also, Van der Pligt, Eiser and Spears, 1986a).

Sample
The sample for this study consisted of 300 residents of the three communities closest to the sites initially proposed by the CEGB (data set 2-*V*; see Figure 2.2). We approached a random sample (*n* = 450) drawn from the electoral registers of the three communities. Of these, 24 had died or moved, and 300 returned the postal questionnaire (a response of 70 per cent). The 11 subjects who responded anonymously were excluded, since the overall design required follow-up questionnaires (see Chapter 2). The average age of respondents was 47.5 years: 24 per cent were younger than 30 years, nearly 46 per cent were between 30 and 60 years of age; and the remaining 30 per cent were older than 60 years of age. The latter percentage reflects the fact that two of the three communities are located in relatively popular retirement areas (30 per cent of this sample were retired). The sample contained 51 per cent males. On average, the respondents had lived 22.9 years in the area and 63 per cent were houseowners.

Questionnaire

The questionnaire contained eight pages preceded by a cover letter explaining the purpose of the study and the independence of the research team. The questionnaire was closed-ended and was preceeded by a short introduction explaining the CEGB announcement about the possible sites for the next nuclear power station in the South West of England. First, subjects' attitudes towards building more nuclear power stations in the United Kingdom, the South-West of England, *and* in their locality were assessed in terms of seven categories ranging from 'very strongly opposed' to 'very strongly in favour'.

Similar questions were asked about other industrial developments in the locality. Respondents were then presented a list of eight general beliefs about nuclear energy. The list contained both pronuclear and antinuclear beliefs (e.g., 'Nuclear energy is the only practical source of energy for the future'; 'Britain should abandon all plans to build any more nuclear power stations'). These beliefs were rated on a 7-point scale ranging from 'very strongly disagree' to 'very strongly agree'. Respondents were asked to indicate their involvement with the issue and how much they cared 'whether a new nuclear power station is built in your neighbourhood'. Responses were on 4-point scales ranging from 'not at all' to 'very much'. Subjects were also asked to indicate whether they had attended any public meetings on the issue.

Next, respondents were presented with a series of questions (selected on the basis of previous research and interviews with local residents) concerning the possible impacts of a new nuclear power station in their locality. First, subjects were presented a list of fifteen *direct* impacts of the construction and operation of a new nuclear power station in their neighbourhood and were asked to indicate how each of these would change life in the neighbourhood for the better or for the worse. Responses to these items (A_1 to A_{15}) were given on a 9-point scale ranging from 'very much for the worse' (1) to 'very much for the better' (9). Only the endpoints of the scale and midpoint ('neither better nor worse') were labelled. After completing this section, subjects were asked to select the five consequences they thought to be the most important. Subjects were then presented with a list of fifteen *indirect* impacts of the construction and operation of a

51

new nuclear power station. These items (B_1 to B_{15}) were worded so as to deal with aspects of local life that might or might not be expected to benefit by such a development. Responses were indicated on a 9-point scale, and subjects were again asked to select the five consequences they thought most important. Finally, subjects were asked to indicate how much importance a public inquiry should attach to five specific aspects of nuclear energy. These aspects were: local environmental impact, political implications of a nuclear energy policy, economic impact, the risks of accidents and pollution and, finally, feasibility of other energy technology. Answers were indicated on a 7-point scale ranging from 'no importance at all' (1) to 'extreme importance' (7).

Results

First, we will focus on the distribution of attitude scores towards nuclear developments in the UK, the South-West and the locality. This distribution confirms the NIMBY-effect discussed earlier in this chapter. In the remaining analyses we will first describe differences in the attitudinal structure of those who oppose both local and national expansion of nuclear energy, those who oppose only the local building of a nuclear power station, and those who are in favour of expansion of nuclear energy both locally and nationally. Next, we will investigate the attitudinal structure and the perceived salience of possible impacts in relation to attitudes towards the local building of a nuclear power station.

Attitudes. Respondents were generally opposed to the construction of a nuclear power station in their neighbourhood. As shown in Table 3.1, approximately 75 per cent of the subjects indicated a negative attitude towards the proposed development, whilst only 15 per cent were in favour. Table 3.1 also shows a much more negative attitude towards the local building of a nuclear power station as compared to such a development elsewhere in the South-West or the UK. We also computed an attitude-index score based on the eight general statements concerning nuclear energy. This scale proved reliable and consistent and the scores showed a normal distribution of attitudes with a marginally antinuclear mean of 28.0 (possible range from 8 to 56).

Table 3.1   Frequency distribution of attitude scores

	Percentage in each attitude category						
Attitude towards	1	2	3	4	5	6	7
More nuclear power stations in the UK	28.9	6.6	16.7	26.5	16.4	1.4	3.5
New nuclear power station in locality	57.8	5.5	11.8	14.5	8.7	0.0	1.7
New nuclear power station in the South-West	34.3	6.6	17.6	26.0	12.8	1.7	1.0

Note: Scale ranging from 1 ('very strongly opposed') to 7 ('very strongly in favour')
Adapted from Van der Pligt, Eiser and Spears, 1986a, p. 6.

This score correlated 0.80 with subjects' attitudes towards building a nuclear power station in the South-West of England, and only 0.60 with attitudes towards the building of a nuclear power station in the neighbourhood. These findings confirm the NIMBY-effect and show that attitudes towards a new nuclear power station in the locality are different from, and more extreme than, attitudes towards nuclear energy in general.

Local vs. general opposition and salience. Next we split the sample into four groups of comparable size on the basis of respondents' attitudes towards the building of a new nuclear power station.

Group PN (pro/neutral) consisted of 73 individuals who were opposed neither to building more nuclear power stations in the UK, not to a new nuclear power station in the locality.
Group LO (locally opposed) included 68 people who were not opposed to building more nuclear power stations in the UK, but who were opposed to a new nuclear power station in the locality.
Group MO (moderately opposed) consisted of 66 individuals who indicated that they were moderately opposed to more nuclear power stations both in the locality and elsewhere in the UK.
Group XO (extremely opposed) included 83 individuals who were strongly opposed to more nuclear power stations both in their neighbourhood and in the UK generally.

There were some demographic differences between the attitude groups. Those who were least antinuclear were more likely to be male and employed, and were less likely to be retired, to own their own home, or to have most of their family living in the area. They tended to be younger and perceived themselves as having greater job mobility. Group *PN* reported being less involved in, and concerned least about, the issue of a new nuclear power station. These results are borne out by the fact that only 19 per cent of group *PN* said that they had attended a public meeting on the issue, compared with 60 per cent, 52 per cent, and 61 per cent of groups *LO*, *MO*, and *XO*, respectively, $\chi^2 (3) = 34.42$, $p < 0.001$. The differences in terms of the thirty impact items are shown in Table 3.2 and are based on a reduced sample of 262 with complete data on all these items.

Table 3.2   Mean ratings of expected impacts by each attitude group

		Attitude group			
	Impact	*PN*	*LO*	*MO*	*XO*
A_1	Excavation for pipelines	3.9	2.0	1.9	1.6
A_2	Construction traffic	3.2	1.4	1.6	1.3
A_3	Road building	5.4	2.5	2.4	2.0
A_4	Conversion of land from agricultural use	3.5	1.8	1.4	1.3
A_5	Noise of construction	3.7	1.8	2.0	1.6
A_6	Workers coming into the area	5.8	2.0	2.7	2.2
A_7	Noise of station in operation	4.4	2.9	2.9	2.3
A_8	General appearance of the power station buildings	3.8	1.4	1.8	1.2
A_9	Area of land fenced off	4.0	1.6	1.9	1.3
A_{10}	Steam or water vapour from station when operating	4.1	2.1	2.2	1.5
A_{11}	Increased security and policing	5.5	3.2	3.0	2.1
A_{12}	Warming of nearby sea water	5.5	3.1	3.3	2.2
A_{13}	Transportation of nuclear waste	3.4	2.0	1.3	1.2
A_{14}	Overhead power cables/pylons	3.3	1.7	1.8	1.5
A_{15}	Overall height of buildings	3.5	1.3	1.7	1.2

Table 3.2 continued

	Impact	Attitude group			
		PN	LO	MO	XO
B_1	Employment opportunities	7.9	5.8	6.5	5.7
B_2	Tidiness of the village	4.8	3.3	3.0	2.7
B_3	Standard of local recreational facilities	6.2	4.4	4.6	3.8
B_4	Social life in the neighbourhood	6.1	4.2	4.0	3.7
B_5	Wildlife	3.4	1.5	1.7	1.4
B_6	Marine environment	4.2	2.4	2.2	2.1
B_7	Farming industry	3.7	2.2	1.9	1.5
B_8	Security of local electricity supplies	6.5	5.3	5.3	4.7
B_9	Health of local inhabitants	4.6	3.6	2.6	1.8
B_{10}	Landscape	3.3	1.2	1.7	1.1
B_{11}	Holiday trade	4.5	2.7	2.7	2.0
B_{12}	Business investment	6.3	4.6	4.6	3.7
B_{13}	Your personal peace of mind	4.7	2.3	1.6	1.2
B_{14}	Standard of local transport and social services	6.4	5.1	5.3	4.5
B_{15}	Standard of shopping facilities	6.0	5.3	4.9	4.7
	Mean	4.7	2.8	2.8	2.3
	n	70	64	59	69

Note: Scale from 'very much for the worse' (1) to 'very much for the better' (9).
All Fs ($df = 3,258$) are significant at $p < 0.001$.
Adapted from Eiser *et al.* (1988), p. 658–659.

While the overall means for *LO* and *MO* are quite similar, a majority of the items show smaller differences between groups *LO* and *XO* than those between groups *MO* and *XO*.

In order to further investigate these group differences on the thirty items, we carried out a step-wise discriminant analysis to find out which items most distinguished the three groups. The first discriminant function, accounting for 85.1 per cent of the shared variance, reflected a clear trend across the four groups. This function could almost entirely be defined in terms of item B_{13} ('personal peace of mind') on which the more antinuclear respondents showed the most pessimism. The second function, accounting for 11.5 per cent of the shared variance, essentially discriminated group LO from the remaining groups. The second function yielded more interesting distinctions between the individual items. High scores on this function (as shown by group LO) were associated particularly with higher scores on items B_9 and A_{13}, but lower scores on items such as B_{10}, A_6, A_5, A_{15}, and B_1. In other words, the locally opposed respondents were distinguishable by their relative lack of concern for what may be thought of as specifically nuclear-related impacts ('health of local inhabitants', 'transportation of nuclear wastes'). On the other hand, they showed a particularly great concern for a number of general aspects of environmental conservation versus disruption ('landscape', 'workers coming into the area', 'noise of construction', 'overall height of buildings'). They also were less convinced of the positive consequences for the locality of economic impacts such as increased employment opportunities.

These different perspectives also showed up in the ratings of importance to be attached to various aspects in the event of a public inquiry. Group LO gave the highest ratings of importance to 'local environmental impact', and the lowest to 'political implications of a nuclear energy policy' (see Table 3.3). We will return to this issue in Chapter 5. Overall, these findings provide information about the structure of attitudes of those who oppose nuclear developments both nationally and locally versus those who oppose only the local construction of a nuclear power station. Before discussing these findings we will first describe the differences between groups of respondents as a function of their attitude towards the local building of a nuclear power station.

Local attitudes and salience of impacts. In this section we will thus focus on differences between groups that differ in their attitudes towards the local building of a nuclear power station. Perceptions of the various possible impacts and their importance will be described for pronuclear, antinuclear and undecided respondents.

Subjects were split into three attitude groups on the basis of their answers to the question whether they were opposed to, or in favour

56

of the building of a nuclear power station in their locality. Table 3.4 gives the ratings of the local, direct impacts A_1 to A_{15}. All items show highly significant differences between the three groups all in the expected direction with antinuclear respondents being more pessimistic about the possible local impacts.

Table 3.3 Importance of impacts in event of a public inquiry: mean ratings by each attitude group

Aspect	Attitude group				
	PN	LO	MO	XO	F(3,274)
Local environmental impact	5.48	6.66	6.15	6.54	9.92***
Political implications of a nuclear energy policy	3.52	3.03	4.13	3.88	3.51*
Economic arguments	4.93	4.57	4.68	4.46	0.83
Risk of nuclear accident and pollution	5.99	5.97	6.11	6.71	4.11**
Feasibility of other energy technology	5.05	5.30	5.68	6.32	7.01***
Mean	4.99	5.10	5.35	5.58	4.37***[a]
n	73	67	62	76	

Note: Scale from *no importance at all* (1) to *extreme importance* (7).
[a] Pillai's multivariate F with $df = 15,816$.
* $p < 0.05$; ** $p < 0.01$; *** $p < 0.001$.
Adapted from Eiser *et al.* (1988), p. 661.

We also conducted a discriminant analysis to find out which of the local impacts A_1 to A_{15} most distinguished the three attitude groups. The results of the stepwise solution revealed that the two most important aspects were 'area of land fenced off' and 'conversion of land from agricultural use'. The next most important aspect was 'workers coming into the area'. Three more aspects added significantly to the discriminant function: 'general appearance of buildings', 'increased security and policing' and 'noise of construction'. The three columns on the right give an indication of the perceived importance attached to the various local consequences. The results show four aspects that were rated very differently. Of the pro subjects, 67 per cent regarded road building an important aspect, whereas only 20 per cent of the anti subjects selected this aspect among the five most expected impacts. A similar difference was obtained concerning the prospect of workers coming into the area (53 per cent of the pros and 18 per cent of the antis). The antinuclear respondents attachted greater importance to the possible conversion of land from agricultural (58 per cent vs. 27 per cent). Finally, the pro subjects attached more importance to 'increased security and policing' than the antis; the former group also regards this change as an improvement to life in the locality.

We also looked at the overall perceived importance of the various impacts irrespective of own attitude. The impact most frequently selected among the five most important was the transportation of nuclear waste; this possible impact was selected by 54 per cent of the respondents and was seen as important by all three attitude groups.

Table 3.4 Perception of direct local impacts and their importance

		Mean Score[a]			Importance[b]		
		Pro Attitude ($n = 30$)	Neutral ($n = 41$)	Anti Attitude ($n = 209$)	Pro Attitude ($n = 30$)	Neutral ($n = 42$)	Anti Attitude ($n = 217$)
	Impact						
A_1	Excavation for pipelines	4.7	3.4	1.8*	20	26	19
A_2	Construction traffic	3.9	2.8	1.4*	27	36	37
A_3	Road Building	6.6	4.8	2.3*	67	33	20*
A_4	Conversion of land from agri-cultural use	4.6	2.8	1.5*	27	45	58*
A_5	Noise of con struction	4.0	3.5	1.8*	13	12	15
A_6	Workers coming into the area	7.0	5.0	2.3*	53	29	18*

Table 3.4 continued

Impact	Mean Score[a]			Importance[b]		
	Pro Attitude ($n = 30$)	Neutral ($n = 41$)	Anti Attitude ($n = 209$)	Pro Attitude ($n = 30$)	Neutral ($n = 42$)	Anti Attitude ($n = 217$)
A_7 Noise of station in operation	4.8	4.3	2.6*	7	12	12
A_8 General appearance of the power station	4.8	3.0	1.4*	40	52	54
A_9 Area of land fenced off	5.2	3.0	1.5*	27	26	29
A_{10} Steam from station when operating	4.7	3.7	1.9*	10	14	25
A_{11} Increased security and policing	6.4	5.1	2.7*	20	19	7**
A_{12} Warming of sea water	6.2	5.0	2.8*	13	7	13
A_{13} Transportation of nuclear waste	4.1	2.9	1.5*	57	45	55
A_{14} Overhead power cables/pylons	4.1	2.5	1.6*	23	36	36
A_{15} Overall height of buildings	4.2	2.9	1.4*	40	33	42

[a] Possible range of scores from 'very much for the worse' (1) to 'very much for the better' (9).
[b] The scores represent the percentage of subjects selecting each factor among the five most important. The columns do not add up to 500 because of the inclusion of subjects who chose fewer than five aspects.
* $p < 0.001$; ** $p < 0.005$ (linear F-test).
Adapted from Van der Pligt, Eiser and Spears, 1986a, p.7.

This finding confirms the outcomes of survey research which indicates that the waste issue is very important to the general public (see Chapter 1). Fifty-three per cent selected 'conversion of land from agricultural use' and 'general appearance of the power station' among the most important.

The mean ratings by the three attitude groups of the fifteen indirect impacts (B_1-B_{15}) consequences of the building and operation of a nuclear power station in their locality also showed substantial differences. Again, we conducted a discriminant analysis to find out which aspects most distinguished the three attitude groups. The results revealed three aspects which had considerable predictive power in separating the three attitude groups. These included the perceived effect of the development on one's 'peace of mind' and the effects on the environment and wildlife.

Table 3.5 summarizes these findings; the various impacts are grouped to make comparisons with the existing literature easier. The most important difference between the three attitude groups corresponds to what Otway, Maurer and Thomas (1978) termed 'psychological risk', while the other two aspects are related to what these authors termed 'environmental and physical risk' (see also Chapter 1).

Table 3.5   Perception of indirect local impacts and their importance

	Mean Score[a]			Importance[b]		
	Pro Attitude ($n = 30$)	Neutral ($n = 41$)	Anti Attitude ($n = 209$)	Pro Attitude ($n = 30$)	Neutral ($n = 42$)	Anti Attitude ($n = 217$)
Economic factors						
Employment opportunities	8.3	7.6	6.0[c]	73	57	15[c]
Business investment	6.7	6.0	4.2	27	28	11
Environmental factors						
Wildlife	4.7	2.6	1.5	40	57	67
Marine Environment	5.1	3.6	2.3	13	17	38
Farming Industry	4.6	3.0	1.9	17	45	56
Landscape	4.3	2.6	1.3	23	50	66
Public health and psychological risks						
Health of local inhabitants	5.0	4.4	2.6	20	29	48
Your personal peace of mind	5.4	4.3	1.7	27	17	47

Table 3.5 continued

	Mean Score[a]			Importance[b]		
	Pro Attitude ($n = 30$)	Neutral ($n = 41$)	Anti Attitude ($n = 209$)	Pro Attitude ($n = 30$)	Neutral ($n = 42$)	Anti Attitude ($n = 217$)
Social factors						
Social life in the neighbourhood	6.9	5.7	3.0	30	7	11
Standard of local transport and local services	6.8	6.3	4.9	40	17	5
Standard of shopping facilities	6.6	5.8	4.9	20	14	4

[a] Possible range of scores from 1 ('consequence will affect life in the neigbourhood very much for the worse') to 9 ('very much for the better').
[b] The scores represent the percentage of subjects selecting each factor among the five most important.
[c] The differences between the three attitude groups were significant in all cases ($p < 0.05$).
Adapted from: Van der Pligt, Eiser and Spears, 1986a, p.9.

Subjects were again asked to select the five impacts they regarded as most important. The results showed very marked differences between the three attitude groups. The most notable difference concerned the possible effects on employment opportunities, 73 per cent of the pros selected this impact among the most important, as compared to only 15 per cent of the antis. Overall, the pro respondents stressed the importance of economic benefits, while the antis stressed safety aspects (both environmental and psychological risks). These differences illustrate the importance of including both beliefs about possible impacts and their salience. To give an example, even though the attitude groups showed only minor differences in their evaluation of the impact of local employment opportunities, a majority of the pros found this aspect important, while only a small minority of the antis regarded this impact as being of importance.

Results of this research also suggest that the major differences between the attitude groups concern the less tangible, more long-term nature of the potential negative outcomes. Moreover, the findings suggest that the perception of the *psychological risks* is the prime determinant of attitude as indicated by the very high correlation (0.79) between this impact and attitude towards the local building of a new nuclear power station.

To test this further we conducted a stepwise multiple regression analysis with subjects' attitude towards the local building of a nuclear power station as a dependent variable and their ratings on all thirty possible impacts ($A_1 - A_{15}$ and $B_1 - B_{15}$) as independent variables. Table 3.6 summarizes the results of this multiple regression analysis. As mentioned before, the predictive power of the personal risk item 'peace of mind' is impressive. The remaining impacts only marginally improved the overall predictive power.

Table 3.6   Multiple regression analysis of attitudes

Impact	Simple r	Multiple r	Change in Multiple r^2
Personal peace of mind	0.79	0.79	0.63
Area of land fenced off	0.66	0.83	0.06
Increased security and policing	0.59	0.84	0.02
Employment opportunities	0.45	0.85	0.01

Adapted from Van der Pligt, Eiser and Spears, 1986a, p.1.

Finally, we looked at the strength of the relation between the various composite scores and the single-scale attitude measure. To do this we simply added the scores in each of the two sets of fifteen impacts. We further computed a composite score only for those impacts in each set of fifteen that subjects individually selected among the five most important. Next, a similar score was calculated on the remaining (less important) impacts. Table 3.7 presents a summary of the correlations between the various composite scores and subjects' attitudes towards building a nuclear power station in their locality.

Table 3.7   Relations between single-scale attitude measure and composite scores

Composite score	Mean score	Correlation with attitude[a]
15 direct impacts (A1–A15)[b]	33.8	0.64
5 most important impacts[c]	7.8	0.60
10 less important impacts[d]	25.9	0.57
15 indirect impacts (B1–B15)[b]	54.3	0.65
5 most important impacts[c]	11.2	0.69
10 less important impacts[c]	43.1	0.45

A high score reflects a more favourable attitude

[a] Attitude towards local nuclear power station, ranging from 'very strongly opposed' (1) to 'very strongly in favour' (7).

[b] Possible range from 15 to 135

[c] Possible range from 5 to 45

[d] Possible range from 10 to 90

Adapted from Van der Pligt, Eiser and Spears, 1986a, p.12.

These results show that the strength of the relation between attitude and composite scores is not significantly reduced if we only consider the the five impacts selected as most important by each individual instead of all fifteen. Results concerning the direct impacts show very similar correlations between attitude and the three composite scores. Composite scores on the basis of the more long term impacts show markedly different correlations with attitude. The highest correlation was obtained for the composite score based on the five important impacts (0.69); the composite score based on the ten less important impacts yielded a correlation of only 0.45. It is worth noting that the attitude groups also showed more marked differences in the perceived importance of the second set of impacts (B_1–B_{15}) than in the importance of the direct impacts (see Tables 3.4 and 3.5). For this reason, one would expect the correlational differences to be more marked for the composite scores based on the long-term consequences. All in all, a conception of attitudes based on individually selected important impacts provides a simple account of the structure of the attitudes of the three groups *without* a reduction in the predictive power.

Conclusions

These findings indicate that different attitude groups base their attitudes on different possible impacts of the building and operation of a nuclear power station. Our findings indicate that public attitudes towards building a nuclear power station in the locality are more extreme and more anti than towards building more nuclear power stations elsewhere. Furthermore, public attitudes towards building nuclear power stations (whether in the locality, elsewhere in the South-West, or in the United Kingdom) were more extreme and more anti than towards nuclear energy in general. These findings show higher levels of opposition than a number of studies conducted in the United States (e.g. Hughey *et al.*, 1985; Sundstrom *et al.*, 1981; Woo and Castore, 1980; see also Chapter 1). Our findings suggest that the public has serious doubts about the feasibility of nuclear energy and prefers to postpone further expansion of the industry, especially when the expansion takes place in one's locality. In the first set of analyses we focused on the NIMBY-effect and attempted to investigate the differences between those in favour of nuclear energy in general and the local building of a nuclear power station, those opposed to the local building only and those opposing both local and national expansion of the nuclear industry. These analyses point to the importance of psychological risks as indicated by the scores on the item 'peace of mind'. Results of the second set of analyses confirm the importance of this factor and revealed a very close relationship between psychological risks and attitude. One reason for this could be that our research concentrated on people living very close to the proposed nuclear power station (all within a 5-mile or 8-km radius). Most other research (see Chapter 1) used much wider areas around proposed nuclear power stations. This proximity to environmental hazards is bound to accentuate the role of psychological risk and also seems the most likely explanation for the higher levels of opposition obtained in our research as compared to most other studies.

The results of our analyses also point to other important issues which could help us to better understand attitudes towards the building and operation of a nuclear power station in the locality. First, our results indicate that the understanding of the positions of the pro, anti, and neutral attitude groups is enhanced if *two types* of information are included — beliefs concerning the possible impacts and the importance or salience of the various impacts. Scores based on subsets of impacts individually selected as important proved as good a predictor of attitudes as scores based on the whole set of impacts. The advantage of this procedure is that it allows us to identify the important aspects underlying the attitudes of the various groups and discard the less relevant responses that play a minor role in attitudinal decision processes. Differences between the various attitude groups were especially marked on the indirect impacts. Overall, the pronuclear respondents stressed the importance of economic benefits, whereas the antis stressed the risk factors (both environmental and psychological).

In summary, there seems to be relatively minor disagreement among the various attitude groups concerning short-term disruptions of life in the neighbourhood (Impacts A_1–A_{15}). The major differences between the groups concern the less tangible, more long-term nature of potential negative impacts. It seems that different perceptions of these long-term costs (both in terms of evaluation and importance) play a crucial role in attitude formation. In the next section we will return to these issues and focus on the effects of 'familiarity' or experience with living near a nuclear power station.

Attitudes, familiarity and salience

In another study we investigated attitudinal structure and salience as a function of the experience of actually living near a nuclear power station (Van der Pligt, Eiser and Spears, 1986a). A number of surveys have compared the acceptance of a nuclear power plant among people living near one with that of people who do not. Other studies have followed local opinion in an area where a nuclear power station was being constructed and becoming operational. As discussed in Chapter 1, there is some support for the idea that familiarity may lead to increased willingness to accept the building of a new nuclear power station (e.g., Thomas and Baillie, 1982;

Warren, 1981; Hughey *et al.*, 1985). Rosa and Freudenburg (1993) mention that research carried out in the mid-1970s (before the TMI accident) showed considerable support for nuclear power plants among people living near the facility. More recent studies, however, indicate that host-community support can no longer be taken for granted. Freudenberg and Baxter (1984) provided a summary of 36 local surveys in host communities. Their research focused on localities in which the residents were aware of the facilities but before the facilities has completed their first six months of operation. Results indicated that host communities, once supportive of nuclear facilities sited in their neighbourhood, have become clearly unsupportive of such facilities. Erikson (1990) also found evidence for a local reluctance to have a nuclear power plant constructed locally.

Some evidence suggests that people actually living near a nuclear power plant for a longer period of time tend to relatively underestimate the risks associated with nuclear energy (Ester, Mindell, Van der Linden and Van der Pligt, 1983). This is most likely to be caused by (accident-free) experience, but is also in accordance with a simple notion of dissonance reduction (Festinger, 1957). Our own findings (Van der Pligt, Eiser and Spears, 1987b) suggest that this could be the case, because the underestimation of the risks of nuclear power stations is most pronounced when it concerns the power station in one's own locality and does not seem to apply to estimates of the risks of nuclear power plants elsewhere. In the study presented in this section we investigated attitudes and beliefs concerning the construction of a nuclear power station in one's own locality. More specifically, we studied the effects of familiarity upon attitudes and upon the perception of the various potential impacts (costs and benefits) of the building and operation of a nuclear power plant. In this research we also tested the notion that a consideration of *both* expected impacts *and* their subjective importance or salience can provide a more complete picture than could be obtained from consideration of either factor alone.

Sample

A random sample ($n = 925$) was drawn from the electoral registers for four communities that were either close to the two existing nuclear power stations near Hinkley Point in Somerset, or close to the three sites that had been shortlisted as possible locations for a new nuclear power station in South-West England. These three were the East Cornwall site (Bugle/Luxulyan), the two West Cornwall sites having been already excluded, and the two Dorset sites (Langton Herring and Winfrith). Note that the analyses to be reported treat the Winfrith residents, conservatively, as a group relatively *un*familiar with nuclear power. It should be remembered that this group lived near to a small nuclear energy research establishment, and so were *somewhat* familiar. However, they did not have direct experience of a local nuclear facility on the scale of that already operating at Hinkley Point. Of this sample, 69 respondents had moved from the area. The three samples included in the analyses were 2-*V*, 3-*W* and 3-*X* (see Chapter 2, Figure 2.2). Thirty-five respondents who returned questionnaires without their name and address were excluded from the analysis. A further 36 were excluded for incomplete responses, leaving 648 respondents. The average age was nearly 48 years; 23 per cent were younger than 30 years, nearly 47 per cent were between 30 and 60 years of age, the remaining 30 per cent being older than 60 years. Approximately 50 per cent of the respondents were male.

Questionnaire

Respondents were presented with the closed-ended questionnaire described earlier in this chapter, which was preceded by a short introduction explaining the CEGB announcements concerning the possible sites for the next nuclear power station in the South-West of England. All respondents in the Hinkley Point sample (sample 3-*X*) received the questionnaire *after* the CEGB had announced that the next nuclear power station for which planning application would be made would be a *third* station at Hinkley Point. Respondents of the remaining three communities were approached before and after this announcement. About 70 per cent of these respondents were approached before the announcement that excluded their communities for the near future (sample 2-*V*) while the remaining 30 per cent (sample 3-*W*) received a questionnaire *after* Hinkley Point was selected and one of the remaining three sites was shortlisted for the next power station (after completion of Hinkley Point C). Whether respondents were approached before or after the announcement is irrelevant to the present analyses and will be ignored. The questionnaire was identical to the one described earlier in this chapter.

Results

We again computed an attitude index score based on the eight general statements concerning nuclear energy (Cronbach's alpha was 0.85). This score showed a normal distribution of attitudes with a marginally antinuclear overall mean of 30.1 (possible range from 8 to 56). This attitude score correlated +0.78 with the (single-scale) attitude towards nuclear energy, as indicated on a 7-point scale ranging from 1 ('very strongly opposed') to 7 ('very strongly in favour'). Subjects familiar with nuclear power stations had a slightly more favourable attitude towards nuclear energy than respondents in the other communities. Mean scores were 32.9 and 28.7, respectively. Responses on the single attitude scale revealed similar differences and showed a slightly more favourable attitude in the Hinkley Point sample. The less familiar subjects were more involved with the issue of a possible new nuclear power station in the South-West than subjects at Hinkley Point; mean scores were 2.3 and 2.0, respectively.

In order to investigate people's perceptions of the various impacts we again compared the two groups with respect to their ratings of two sets of fifteen possible impacts. Table 3.8 shows the mean ratings by the high and low familiarity subjects of the fifteen direct impacts ($A_1 - A_{15}$) of the building and operation of a nuclear power station in the locality. Subjects were split into two groups on the basis of their location; one group of respondents who lived within 5 miles from the existing nuclear power station at Hinkley Point ($n = 218$), and those who lived in the communities that were shortlisted by the CEGB as a possible future location for a new nuclear power station ($n = 430$). Results in Table 3.8 show substantial differences for all impacts. Overall, the less familiar respondents are more pessimistic about the immediate impacts. The results reported in Table 3.8 are based on a one-way analysis of variance. Since the two groups also differed in their attitude towards nuclear energy, we conducted analyses of covariance with familiarity as an independent variable and the attitude-index score as a covariate. Results of these ANCOVA's showed that attitude was significantly related to the perceived impacts. All effects due to familiarity, however, remained significant although they were less pronounced than the effects reported in Table 3.8.

Table 3.8 also indicates the effect to which each impact was selected among the (five) most important by the two attitude-groups. We conducted a discriminant analysis to find out which aspects most distinguished the two groups of subjects. The results of the stepwise solution revealed six impacts that discriminated most between two groups. The first aspect was 'workers coming into the area' which was seen as far more important by the more familiar respondents. The next most discriminating item was 'overall height of the buildings', regarded as more important by the less familiar respondents. The next items of importance were 'excavation for pipelines' and 'steam from station when operating', both seen as more important by the respondents less familiar with a nuclear power plant. Two more aspects added significantly to the function 'road building' and 'area of land fenced off'. Again, both were seen as more important by the less familiar respondents.

Table 3.9 shows the mean ratings of the fifteen indirect local impacts (B_1–B_{15}) of the building and operation of a nuclear power station in the locality. Results confirm those presented in Table 3.8 and indicate a more pessimistic perception of the various impacts by the less familiar respondents. The most striking differences were found for the expected employment opportunities and consequences for the environment and public health. The Hinkley Point sample is more optimistic about employment opportunities and less pessimistic about the impact on the environment and public health. In accordance with the previous sections, the less familiar respondents were more concerned about their 'peace of mind' than those living near the existing nuclear power stations. Closer inspection of the overall mean scores reveals that both groups, on average, think that most factors will have a negative impact on life in the locality (the majority of the mean scores are lower than five).

Table 3.8 Perception of direct impacts as a function of familiarity

		Mean Score[a]		Importance[b]	
	Impact	High familiarity ($n=218$)	Low familiarity ($n=430$)	High familiarity ($n=218$)	Low familiarity ($n=430$)
A_1	Excavation for pipelines	4.3	2.6*	6	19*
A_2	Construction traffic	3.1	2.0*	42	33*
A_3	Road building	5.1	3.4*	18	24
A_4	Conversion of land from agricultural use	3.4	2.1*	37	50*
A_5	Noise of construction	3.9	2.4*	11	13
A_6	Workers coming into the area	4.8	3.7*	52	22*
A_7	Noise of station in operation	4.3	3.2*	13	12
A_8	General appearance of the power station	3.9	2.2*	32	48*
A_9	Area of land fenced off	4.0	2.3*	19	26*
A_{10}	Steam from station when operating	4.2	2.6*	7	21*
A_{11}	Increased security and policing	5.2	3.7*	15	11
A_{12}	Warming of sea water	5.0	3.7*	14	12
A_{13}	Transportation of nuclear waste	3.1	1.9*	49	55
A_{14}	Overhead power cables/pylons	3.0	2.0*	40	37
A_{15}	Overall height of buildings	3.9	2.1*	15	38*

[a] Possible range of scores from 1 ('very much for the worse') to 9 ('very much for the better').
[b] The scores represent the percentage of subjects selecting each factor among the five most important. The columns do no add up to 500 because of the inclusion of subjects who chose fewer than five aspects.
* Differences between groups significant at $p < 0.05$.
Adapted from Van der Pligt, Eiser and Spears, 1986b, p.82.

Table 3.9   Perception of indirect impacts as a function of familiarity

	Impact	Mean Score[a]		Importance[b]	
		High familiarity (n=218)	Low familiarity (n=430)	High familiarity (n=218)	Low familiarity (n=430)
B_1	Employment opportunities	8.2	6.6*	61	31*
B_2	Tidiness of the village	4.4	3.6*	8	13
B_3	Standard of local re-creational facilities	5.3	4.9*	7	8
B_4	Social life in the neighbourhood	5.1	4.7*	17	10*
B_5	Wildlife	3.6	2.2*	33	57*
B_6	Marine environment	4.1	3.0*	23	28
B_7	Farming industry	3.8	2.5*	30	45*
B_8	Security of local electricity supplies	5.5	5.6	19	9*
B_9	Health of local inhabitants	4.2	3.2	36	40
B_{10}	Landscape	3.4	2.0*	31	53*
B_{11}	Holiday trade	4.8	3.2*	7	18*
B_{12}	Business investment	6.3	5.0*	22	15*
B_{13}	Your personal peace of mind	3.9	2.7*	33	35
B_{14}	Standard of local transport and social services	5.5	5.5	20	10*
B_{15}	Standard of shopping facilities	5.4	5.4	17	1*

[a] Possible range of scores from 1 ('very much for the worse') to 9 ('very much for the better').
[b] The scores represent the percentage of subjects selecting each factor among the five most important. The columns do no add up to 500 because of the inclusion of subjects who chose fewer than five aspects.
* Differences between groups significant at $p < 0.05$.
Adapted from Van der Pligt, Eiser and Spears, 1986b, p.85.

Again, we conducted a discriminant analysis to find out which possible impacts most distinguished the two groups. This revealed four impacts which had considerable predictive power in separating the two groups. The first impact concerned employment opportunities (seen as a positive impact by the more familiar respondents). Further significant contributions were made by three impacts seen as more negative by the less familiar respondents, i.e. 'wildlife', 'holiday trade' and 'landscape'.

Next, we will take a closer look at the contribution of perceived importance or salience to the understanding of the attitudinal differences between the two groups. Inspection of Tables 3.8 and 3.9 suggests that the inclusion of both the perception of the various impacts *and* the perceived importance attached to each of these impacts provides a more complete picture of the attitudinal differences between the two groups.

Table 3.8 indicates that the less familiar subjects were more pessimistic about the immediate impact of a nuclear power station. More specifically, these subjects were more pessimistic about workers coming into the area and the appearance of the buildings. They also find the latter aspect more important than the respondents at Hinkley Point. Table 3.9 indicates that the less familiar respondents see the various risks of nuclear power stations as both 'more serious' and 'more important' than the Hinkley Point sample. This group is not only more optimistic about the economic impact, but also attaches greater value to the possible economic benefits.

Finally, we conducted a stepwise multiple regression analysis with attitude towards nuclear energy (measured by the eight-item attitude scale) as a dependent variable and respondents' ratings of all impacts (A_1–A_{15} and B_1–B_{15}) as predictors. Results are summarized in Table 3.10.

Table 3.10   Multiple regressions analysis of attitudes

Impact	Simple r	Multiple r	Change in Multiple r^2
Personal peace of mind	0.62	0.62	0.38
Transportation of nuclear waste	0.53	0.65	0.05
Employment opportunities	0.42	0.68	0.03
Health of local inhabitants	0.58	0.70	0.03

Adapted from Van der Pligt, Eiser and Spears (1986b), p. 86.

72

These findings show again that the impact on 'peace of mind' and the nuclear waste issue are most predictive in relation to subjects' attitude to nuclear energy and indicate that the risks of nuclear energy are the prime determinant of attitudes to nuclear energy, with people's perception of the economic benefits playing a slightly less important role.

Conclusions

The results of the present study show that attitudes towards the building of a new nuclear power station in one's locality are a function of its perceived impacts and the importance attached to these impacts.

Findings show a marginally more favourable attitude towards nuclear energy in general and towards the building of a nuclear power station in one's locality for people who are presently living near a nuclear power station. Those living near a nuclear power station were less pessimistic about the immediate impacts (e.g. 'workers coming into the area', 'height of the buildings', 'excavation for pipelines' etc...). Familiarity also affected expectations and evaluations of the indirect impacts. People living near a nuclear power station were more optimistic about the impact of employment opportunities and found this factor more important than those not living near a nuclear power station. The latter were more pessimistic about the risks for public health and the environment, and also attached greater value to safety related issues.

Although the Hinkley Point sample showed a slightly less unfavourable attitude towards the building of a new nuclear power station, our results indicate that the overall attitude is unfavourable. Respondents, irrespective of experience of living near a nuclear power station, expect the building of a new nuclear station to have a negative impact on life in their locality. These findings are in accordance with those of Hughey, Sundstrom and Lounsbury (1985; see also Chapter 1). Overall, our respondents' perceptions of the various direct and indirect local impacts were unfavourable. One of the few exceptions concern economic impacts such as employment opportunities which are rated as more favourable by those who live near a nuclear power station.

Impacts differ in their salience to different groups. Attitudes towards local versus national developments of nuclear power, attitudes towards the local building of a new nuclear power station and familiarity with living near a nuclear power station all affect the perceived salience of the possible outcomes of a new nuclear power station in the locality. Differences in salience of outcomes are also related to the NIMBY-effect. In the next chapter we will return to that issue and compare attitudes towards the local building of a new nuclear power station with those towards other industrial developments.

Chapter 4

Comparison with Other Energy Sources

Context and framing

All attitudes are relative. To say that one is in favour of something is to imply that one approves of it more than of something else. This 'something else', however, often is not mentioned and may not be specifically thought about. Approval implies preference, so in order to understand patterns of approval or disapproval, we need to understand what is being preferred to what. What is the context of such preference and what are the standards of comparison? When local residents are asked about their approval of plans for a new nuclear power station, what comparisons will they make? Will they think of a new nuclear power station locally in comparison to the prospect of nuclear power stations elsewhere, in comparison to other possible or hypothetical local developments that are *not* nuclear, or in comparison to the absence of any development at all?

From the point of view of the CEGB, under a statutory obligation to provide secure electricity supplies, proposals for nuclear development imply a preference for nuclear power generation over other methods of generating electricity (or at least over a national energy policy from which nuclear power was excluded). The reasonableness of such a preference is not our concern here. The point is that the CEGB's advocacy of nuclear power was based on a 'means-end' argument within a context set by a non-negotiable end (secure electricity generation) and by a limited set of alternative means (other sources of energy). The CEGB's preference for the nuclear option can be defined as a choice that was at least partly constrained or 'framed' by the context in which decisions had to be made. This chapter is concerned with how the framing of decision contexts can influence choice, preference and attitude.

Different contexts yield different choices and there is always more than one context in which a given option can be viewed. The context in which the CEGB made its choice would not necessarily (or even probably) correspond exactly with that within which villagers considered the prospect of local nuclear development. Security of national electricity supply was not the responsibility of local residents, nor the primary end with which they would be concerned. Another way of expressing this is to say that different aspects of the issue were salient to the industry on the one hand and to local residents on the other (as well as to different groups of residents), as the findings presented in Chapter 3 illustrate. A divergence between the preferences of the industry and those of local residents, therefore, could follow naturally from the different contexts in which these preferences were formulated.

Oil and nuclear developments

So far, our research has been almost exclusively concerned with nuclear energy. This was a consequence both of the places and times of our surveys, and of the structure of our questionnaires as a whole, which contained many sections dealing with nuclear power, besides being introduced with reference to the CEGB's proposals. Considering the 'four counties' study as a whole, no balance was attempted, let alone achieved, between a focus on nuclear energy and on any other single technology. Furthermore, even though our respondents were ready (as we shall see later in this chapter) to express attitudes about non-nuclear technologies and developments, these were hypothetical prospects compared with the proposal for a new nuclear power station near their homes.

Some three years later, however, an opportunity arose to compare the attitudes of local residents to two separate forms of energy development in their neighbourhood. As will be remembered, the CEGB's final announcement of August 1982, while identifying Hinkley Point as the first choice for development, explicitly mentioned Winfrith in Dorset as a likely site for a new power station in the next round of expansion of nuclear generating capacity. Although Winfrith already had a small research reactor and although

76

the proposed development would be largely confined to the land already occupied by the existing plant, this would constitute a major expansion of the local nuclear presence.

Nuclear power, however, was not the only energy technology with special relevance to this region. Only about 10 miles (16 km) to the east of the Winfrith station is the Wytch Farm oil field owned by British Petroleum (BP). Exploratory work at Wytch Farm started in 1970/1 and the first well was sunk in 1974. By the end of the decade, Wytch Farm had become the largest onshore oil development in the United Kingdom, with production rising to around 4,000 barrels a day. (All this would be difficult to guess from the modest visible signs of the operation, largely hidden from view by surrounding woods.) Continued exploration in the early 1980s revealed signs of a further oil reservoir, deeper and larger than that already known, extending from close to Wytch Farm out under the sea in the bay known as Poole Harbour. To assess the scale of this find, BP applied for permission from Dorset County Council to drill appraisal wells on Furzey Island — a small wooded island in Poole Harbour just to the north of Wytch Farm. This application was approved on 12 April 1985. We were thus provided with an opportunity to compare the attitudes of local residents to existing and proposed future developments of two different energy industries, within an area of considerable environmental interest and beauty.

The Dorset 'proximity' study

The first study we designed to exploit this opportunity (see Eiser, Spears, Webley and Van der Pligt, 1988) was concerned mainly with looking for any influence due to the relative proximity of the oil and nuclear developments to people's homes. We also attempted to identify which aspects of either development were most predictive of overall attitudes. These issues relate closely to those considered in Chapter 3 with specific reference to nuclear power. How does previous experience or familiarity with a particular technology influence the attitudes of local residents, and which aspects of any proposed development make the most salient contribution to such

77

attitudes? To address these issues, we designed a questionnaire including sections based closely on those used in Chapter 3 studies.

This study was conducted in April 1985, immediately after BP obtained drilling permission. It involved the distribution of 800 questionnaires to randomly chosen names from the electoral registers of four communities or areas (200 each) around the town of Wareham. Of the 800, 42 had left their addresses; 402 (53.0 per cent) of the remainder responded immediately or after one reminder. The four communities broadly comprised four quarters of a geographic square. Specifically these were: (i) in the south-west, Winfrith and Lulworth; (ii) in the north-west, Wool; (iii) in the south-east, Corfe and Arne; and (iv) in the north-east, part of Poole (mainly the Sandbanks peninsula). These were treated as a 2 x 2 factorial design, in that the two more westerly communities are closer to the nuclear development, whereas the other, more easterly, communities are closer to the oil developments. At the same time, the two more northerly developments are more heavily built up. The two more southerly communities are more rural, consisting of smaller villages and hamlets and each containing well-known tourist attractions (Lulworth Cove and Corfe Castle).

The questionnaire was designed to measure respondents' attitudes towards the following oil developments: the existing oil wells around Wytch Farm, new oil wells around Wytch Farm, new oil wells along the south coast (not Dorset), and new oil wells in the North Sea. The following nuclear developments were also rated: the existing nuclear power station at Winfrith, a new nuclear power station at Winfrith, a new nuclear power station in North Somerset, and a new nuclear power station elsewhere in the UK (not Dorset or Somerset). Responses were in terms of a 9-point scale from 'extremely opposed' (1) to 'extremely in favour' (9).

There then followed four pages, each containing a list of twelve aspects of 'life in your neighbourhood'. Respondents had to judge how each of these might be changed (in terms of a 9-point scale from 1 = 'very much for the worse' to 9 = 'very much for the better ') by (a) the drilling of new oil wells, (b) the operation of new oil wells, (c) the construction of a new nuclear power station, and (d) the operation of a new nuclear power station. The twelve aspects were the same on all four pages, and were as follows: peace and quiet;

freedom to come and go as you please; effects on agriculture; your personal health and safety; your personal financial situation; local business/employment opportunities; environmental pollution; the standard of local amenities and services; visual impact; the character of your town/village; local tourist trade; and the risk of a major environmental disaster.

The mean attitudes expressed by the four communities towards existing and possible new oil and nuclear developments are shown in Table 4.1. As far as attitudes to the existing oil and nuclear developments were concerned, the oil wells at Wytch Farm were rated more positively than the nuclear power station at Winfrith ($p <$ 0.001), irrespective of how near to either development residents lived. As regards the six possible new developments, oil was strongly preferred to nuclear power ($p < 0.001$). The strongest effects of proximity of residence on attitude are found when the specific type and place of development is taken into account. As far as oil was concerned, those living nearer Wytch Farm approved more of new oil wells, either locally or elsewhere on the south coast (means respectively: 6.6, 6.1) than did those living nearer Winfrith (5.6, 5.3). In the case of nuclear power, however, the more easterly communities showed less differentiation between a new plant at Winfrith (3.8) compared with one at Hinkley Point (4.4) than did those living nearer Winfrith (3.6, 4.8, respectively). As expected from the findings reported in Chapter 3, there was a general preference for a new nuclear power station to be further away from one's home, and this was especially true among those living nearest to the proposed site. This was not the case for new oil wells, where a more local development was preferred, especially by those living closest to the exploration.

Ratings of anticipated change to neighbourhood life also showed group differences. The overall impact of nuclear development was evaluated more negatively than that of oil ($p < 0.001$), especially in the more rural southerly communities and among those living nearer to Winfrith than to Wytch Farm. The operation phase of either development was viewed more pessimistically than drilling or construction ($p < 0.01$), this being rather clearer in the case of nuclear power and among the more rural groups. Considering separate aspects, those where nuclear power compared particularly badly with oil were personal health and safety, and the risk of a major disaster.

Table 4.1   Attitudes to oil and nuclear developments: 'proximity' study

| | Town (north) | | Village (south) | | |
	Nuclear (west)	Oil (east)	Nuclear (west)	Oil (east)	Overall
Oil					
Existing oil wells	7.2	6.1	6.7	7.3	6.8
New oil wells locally	5.9	5.2	6.3	6.1	6.1
Elsewhere in region (south coast)	6.3	5.2	5.8	5.4	5.7
Elsewhere in UK (North Sea)	7.9	8.2	8.2	7.8	8.0
Nuclear					
Existing nuclear station	5.7	4.8	5.0	4.7	5.1
New nuclear station locally	4.2	4.1	3.0	3.6	3.7
Elsewhere in region (North Somerset)	4.9	4.6	4.7	4.1	4.6
Elsewhere in UK	5.5	5.2	5.2	4.8	5.2

Note: Possible range of scores from 1 ('extremely opposed') to 9 ('extremely in favour').

We next performed a number of analyses which involved classifying respondents, regardless of locality, into different groups depending on their attitudes to nuclear and to oil developments (as expressed on the items shown in Table 4.1). Classification of nuclear attitudes was made on the basis of essentially the same criterion as that used in the analyses reported in Chapter 3 (except that 'extremely' and 'moderately opposed' positions were combined): that is, there was an *opposed* group consisting of any individuals who said they were opposed to a new power station both at Winfrith and elsewhere in the UK, a *locally opposed* group who objected to a new power station at Winfrith but not elsewhere in the UK, and a *neutral/pro* group who did not object to a new power station at Winfrith (or elsewhere).

The different nuclear attitude groups were then compared with respect to their ratings of anticipated impacts. Considering both the drilling/construction and operation phases, the opposed group held the most pessimistic, and the neutral/pro group the most optimistic expectations of possible impacts. On almost all items (the exceptions being business/employment during both phases and disaster risk during construction) the locally opposed group occupied a position close to that of the opposed group, rather than one midway between the other two. Further analyses suggested that concerns about pollution and health and safety (during operation), disaster risk (during both phases) and less optimism about business/employment (during construction) were especially important in distinguishing the opposed from the locally opposed group. Concerns about visual impact and (during operation) character of the town/village contributed most to differentiating the locally opposed from the neutral/pro group.

Attitudes to oil, being far more favourable, permitted the following tripartite classification: a *neutral/anti* group who did not support new oil wells either at Wytch Farm or elsewhere on the south coast, a *pro* group who favoured new oil wells either at Wytch Farm or elsewhere on the south coast, and a *very pro* group who strongly supported new oil wells both at Wytch Farm or elsewhere on the south coast (with rating of 8 or 9 on the 9-point scale). We then compared the three oil attitude groups in terms of their ratings of the separate impacts of drilling and operation of new oil wells. The main finding was a simple association of attitude with ratings of anticipated consequences. That is, the neutral/anti group gave the most negative, and the very pro group the most positive ratings of all anticipated impacts within both the drilling and operation phases.

Another important feature of these data is that attitudes to nuclear and oil developments were positively associated ($p < 0.001$). Thus, there is evidence of attitude organization at both more specific and more general levels. On the one hand, those opposing a given development held negative expectations about its local consequences. This is consistent with a view of attitudes as involving structures of 'expectancy-value' beliefs. On the other hand (although there was a

81

clear preference for oil over nuclear development), some of this opposition seemed to be based on a generalized objection to any form of industrial expansion.

Direct and indirect aspects of oil and nuclear developments

A further study (see Eiser, Spears and Webley, 1988) comparing attitudes to the proposed developments at Wytch Farm and Winfrith was conducted a year later (April 1986). This involved a similar procedure, with a focus on the aspects of each development anticipated to change life in the neighbourhood for better or worse. The aspects, worded as far as possible in the same way for the two developments, were chosen so as to reintroduce the distinction used in Chapter 3 between 'direct' (more immediate or certain) and 'indirect' (longer term or conjectural) effects, while no longer attempting to compare directly the impacts of the construction and drilling phases. Attitude groups were also distinguished in a manner more directly comparable to that used in Chapter 3.

The sample consisted of 480 residents of Wareham and its immediate surroundings. This constituted a 52.6 per cent response rate from a total mailing of 1033, reduced by 120 no longer at their registered addresses. Of the 469 who declared their sex, 290 were men and 179 were women. As in our previous studies, the final sample consisted of all those responding either immediately, or after a single reminder. This has a special significance for this study, since the reminder was dispatched on April 21st, after 356 replies had been received. The Chernobyl accident occurred on April 26th, and it is possible (but not certain in all cases) that later respondents would have heard about this before responding. In fact, later respondents expressed more antinuclear attitudes, but the design of this study does not allow us to distinguish between the possible impacts of the accident and of self-selection. We shall return to this question in Chapter 6.

Attitudes to oil were classified on the basis of expressed approval of new oil wells at Wytch Farm and elsewhere inshore in the UK. Four groups were distinguished: very pro, who declared themselves 'strongly' or 'very strongly in favour' of new oil wells both locally and nationally; moderately pro, who were 'in favour' of both options; locally neutral/anti, who were opposed or neutral towards new oil wells locally and who gave a more positive rating of new oil wells nationally; and neutral/anti, who did not approve of new oil wells locally or nationally, and who indicated no preference for the national over the local option.

Nuclear attitudes were similarly classified into four groups: neutral/pro, who did not oppose a new power station, either at Winfrith or elsewhere in the UK; locally opposed, who opposed a new station locally but not elsewhere; moderately opposed, who were 'opposed' or 'strongly opposed' to both options; and extremely opposed, who were 'very strongly opposed' to both options. Table 4.2 shows the cross-tabulation of the frequencies in the two sets of attitude groups, which were positively associated ($p < 0.001$).

Table 4.2   Association between oil and nuclear attitudes: 1986 Dorset study

| Attitudes to nuclear power | Attitudes to Oil | | | | |
	Very Pro	Moderately Pro	Neutral/ Anti	Locally Anti	Total
Neutral/Pro	33 (24.1%)	65 (47.4%)	15 (10.9%)	24 (17.5%)	137
Locally Opposed	7 (11.3%)	22 (35.5%)	25 (40.3%)	8 (12.9%)	62
Moderately Opposed	5 (4.8%)	39 (37.5%)	33 (31.7%)	27 (26.0%)	104
Extremely Opposed	10 (9.3%)	29 (26.9%)	25 (23.1%)	44 (40.7%)	108
Total	55	155	98	103	411

Note: Percentages are of row totals.

Table 4.3 shows the mean ratings given by the four oil attitude groups to the seventeen aspects/impacts, when asked to say how each would change life locally, or be changed by new oil wells locally (in terms of the same 9-point scale as before, from 'very much for the worse' to 'very much for the better').

Table 4.3   Perceived impacts of new oil wells
             Means for attitude groups and univariate Fs: 1986 Dorset
             study

Impact	Attitudes to Oil				
	Very Pro	Moderately Pro	Locally Neutral/Anti	Neutral/ Anti	F (3,383)
Noise of construction/ operation	4.7	4.3	3.0	2.8	29.99
Conversion of undeveloped land	5.3	4.6	2.7	2.5	43.65
Excavation for pipelines	5.3	4.4	2.7	2.6	42.97
Workers coming into the area	5.3	4.6	3.5	3.5	13.08
Area of land fenced off	4.6	4.1	2.5	2.8	28.25
Visibility of new buildings	4.8	4.0	2.7	2.8	30.74
Visibility of drilling rigs	4.6	4.0	2.5	2.5	39.14
Transportation/storage of oil	4.8	4.0	2.6	2.6	32.83
Road building	6.8	5.2	3.8	3.1	41.07
Employment opportunities	7.4	6.7	5.9	5.7	14.08
Prosperity of your town/ village	7.1	6.2	5.6	5.2	17.06
Standard of local facilities/services	6.3	5.8	5.1	4.8	14.97
Wildlife	4.5	3.7	2.6	2.4	25.47
Health of local inhabitants	5.1	4.9	4.0	4.0	14.20
Landscape and marine environment	4.9	4.0	2.6	2.7	35.12
Risk of a major accident	4.6	4.0	3.1	2.7	24.97
Tranquillity of daily life	4.8	4.4	3.1	2.9	35.36
$n =$	48	149	93	97	387

Note: Scores coud range from 1 ('very much for the worse') to 9 ('very much for the better').
All Fs are significant at $p < 0.001$.

Table 4.4   Perceived impact of new nuclear power station
Means for attitude groups and univariate Fs: 1986 Dorset
study

Impact	Attitudes to Nuclear Power				
	Neutral/ Pro	Locally Opposed	Moderately Opposed	Extremely Opposed	F (3,42)
Noise of construction/ operation	4.5	3.0	3.2	2.3	44.23
Conversion of undeveloped land	4.9	3.2	2.9	1.7	61.09
Excavation for pipelines	4.9	3.6	3.4	2.3	48.41
Workers coming into the area	5.1	3.8	3.9	3.2	17.71
Area of land fenced off	4.5	3.1	3.2	2.0	51.25
Visibility of new buildings	4.2	2.5	2.6	1.9	51.88
Visibility of drilling rigs	3.7	2.0	2.5	2.1	29.45
Transportation/storage of radioactive materials	3.4	1.8	1.4	1.1	77.77
Road building	5.7	4.6	4.1	3.4	21.04
Employment opportunities	7.4	6.4	5.9	5.8	23.10
Prosperity of your town/ village	6.8	5.9	5.3	4.9	24.83
Standard of local facilities/services	6.2	5.2	5.2	4.6	20.98
Wildlife	4.0	2.3	2.6	1.7	46.55
Health of local inhabitants	4.6	3.4	2.6	1.6	96.89
Landscape and marine environment	4.4	2.7	2.4	1.5	88.93
Risk of a major accident	4.0	2.6	2.0	1.4	95.58
Tranquillity of daily life	4.5	3.2	3.0	2.4	46.93
$n=$	147	64	103	110	424

Note: Scores coud range from 1 ('very much for the worse') to 9 ('very much for
the better').
All Fs are significant at $p < 0.001$.

As can be seen, the means follow an order extremely consistent with overall favourability. There is a clear split between the two favourable and the two neutral/anti groups, with the means for the latter two being quite close together.

The means for the nuclear ratings are shown in Table 4.4. It is worth noting that seven impacts, mainly concerned with local nuisance and visual impact, show the locally opposed group lying between the moderately opposed and extremely opposed groups. On the other hand, items concerning local economic benefits and health and accident risks tend to show the locally opposed group as rather less pessimistic than the moderate opponents.

The findings confirm our previous conclusions regarding attitudes to the prospect of a new nuclear power station, namely that much local opposition rests at least partly on concerns that are not specifically antinuclear in origin. The values placed on conservation of natural wildlife and environment, as well as the ambience and appearance of the neighbourhood, constitute powerful motives for resisting major development of any kind. Health and safety issues, while important concerns for all, are regarded particularly fearfully by those who oppose nuclear development in a national, rather than just a local, context. Attitudes to oil development are generally far more favourable, but here again, the withholding of unreserved approval seems to reflect a similar protective concern.

This parallelism between oil and nuclear attitudes is extremely important. If we attempt to understand the attitudes of local residents by asking 'What is it specifically about nuclear power that people don't like?' we might get one set of answers. If we ask the same question about oil, we might get another set. There may be little in the answers to explain why those people more opposed to oil would also object to nuclear power. If, however, we ask a question such as 'What is it specifically about the local environment that residents most value and wish to preserve?' the specific form of any industrial development, be it oil or nuclear or something else, may be somewhat secondary. Where we find different attitudes, there we will also find different frames of reference. The frame of reference for developers involves a focus on something which does not yet exist but which they would like to build. The frame of reference for local residents more frequently involves a focus on what does exist but which they do not wish to lose.

Combinations of prospects

The question 'What are people comparing with what?' has more general relevance for attempts to understand public attitudes towards nuclear and other developments. The traditional 'expectancy-value' approach to decision-making — incorporated into the 'subjective expected utility' (SEU) theory of economic rationality as well as some models of attitude formation, notably the 'theory of reasoned action' (Fishbein, 1967; Ajzen and Fishbein, 1980) — assumes that different prospects are evaluated in terms of the end-states they are predicted to produce. An important departure from this tradition has been put forward by Kahneman and Tversky (1979) under the title of 'prospect theory'. The central tenet of prospect theory is that expected consequences are not evaluated as end-states in absolute terms, but as changes from some comparison standard, which could be some future target or the existing status quo.

Prospect theory has inspired a series of experimental demonstrations of different preferences for prospects of comparable expected value, depending on whether decision problems are 'framed' in terms of gains or losses relative to some reference point (Fischhoff, 1983; Meyerowitz and Chaiken, 1987). A finding predicted by Kahneman and Tversky is that individuals will tend to avoid risk when faced with what appears to be a choice between gains of different size and uncertainty, but will be more prepared to take risks to avoid accepting a certain loss (see Eiser and Van der Pligt, 1988, for a more general discussion). Our research comparing attitudes to oil and nuclear developments was not concerned directly with this distinction between gains and losses. However, prospect theory might imply that levels of acceptance of either one of the two developments would be contingent on assumptions about the likelihood of the other development proceeding. In other words, would residents feel differently about either proposal if it were the only development going ahead, than if they felt they were having to consider a future in which both developments were proceeding? A strict SEU approach would have some difficulty in explaining any impact of different reference points on attitudes to either development considered singly.

An attempt to consider this question was made at the same time as the first ('proximity') study (see Eiser, Spears, Webley and Van der Pligt, 1988) involving 421 respondents to a questionnaire mailed to 835 residents of the Dorset town of Wareham, halfway between Winfrith and Wytch Farm. The design involved asking respondents to rate their own attitudes towards the nuclear and/or the oil development, and also to estimate the likelihood that the development would go ahead, with different wordings of the questions. Basically, these questions asked either for simple ratings of each development by itself, or for responses based on the assumption that the other development would or would not proceed. For example, the question 'How would you feel about a new nuclear power station at Winfrith?' was presented both by itself ('unframed' condition) and with the riders: (i) 'if there were also definitely going to be new oil wells around Wytch Farm?' and (ii) 'if there were also definitely going to be no new oil wells around Wytch Farm?' ('framed' condition).

There was an overall preference for oil over nuclear power (as one would expect from the findings reported earlier), and more positive attitudes towards the existing oil wells and/or power station were associated with greater acceptance of possible expansion of the relevant industries. However, the effects of 'framing' the questions so as to introduce different reference points were mixed. Attitudes towards new oil wells were significantly ($p < 0.001$) less positive when respondents were asked to assume that there would also be a new power station at Winfrith. Also, those who approved of the oil development also considered it somewhat more likely ($p < 0.05$). However, consideration of the oil wells had no consistent impact on how the nuclear development was rated. This asymmetry is probably attributable to the fact that the oil development, considered by itself, elicited generally positive attitudes, and therefore would not necessarily be seen as adding to the negative consequences of a new nuclear power station. Despite this, we believe that more attention should be paid to how different hazards are perceived in combination, since many sources of pollution and health risks occur together with each other, so that their impacts can typically be cumulative.

Alternative energy technologies

The Dorset residents in the studies just described were in the comparatively rare situation of being faced with two different kinds of energy developments in their neighbourhood, neither of which directly competed with the other. In a national and international context, however, nuclear and other technologies are typically presented as competing alternatives. Opposition to nuclear power can indeed involve questioning the assumption that more electricity generating capacity is needed at all, but more commonly it rests on the argument that this capacity can be supplied by other means. How, then, were the relative merits of nuclear power compared with other energy sources, viewed by local residents of communities in which a new nuclear power station might be built?

Returning to the data from our earlier 'four counties' study, we have already noted that much local opposition to the CEGB's proposals was not specifically an opposition to nuclear power as such. As we saw in Chapter 3, many of our sample objected to a nuclear power station in their own neighbourhood but not elsewhere in the country. Their reasons for doing so seemed to reflect a concern for aspects of their immediate environment and of the character of their community which would be changed by virtually any large-scale industrial development. The irony is that the attributes of the various sites (such as their remoteness from large centres of population) that rendered them plausible locations for nuclear development meant that other kinds of development were comparatively unlikely.

As described in Chapter 3, some of the respondents in our 'four counties' study (sample 2-V) were asked to express their levels of approval of nuclear developments locally and nationally. These same respondents, however, evaluated four hypothetical local developments. These were: a coal-fired power station, windmills for generating electricity, a chemicals factory, any industrial development. All were followed by the phrase 'taking up the same area of land in your neighbourhood'. Table 4.5 shows the percentages expressing different levels of approval of the different prospects. As can be seen, much of the expressed antinuclear feeling was not distinctively antinuclear. Only 2.7 per cent of the sample expressed any support for the idea of a local chemicals plant, with 90.3 per cent opposed. Only windmills achieved a rough balance between support and opposition (with 9.9 per cent declaring themselves to be very strongly in favour). Opposition to a coal-fired power station in the neighbourhood was about the same as opposition to a new nuclear power station elsewhere in the South-West region. While these data further illustrate the kind of NIMBY phenomenon described by previous researchers (see Chapter 1), they reinforce the message that nuclear attitudes cannot be interpreted in isolation.

Table 4.5   Attitudes to new nuclear power stations and hypothetical local developments (percentages responding in each category)

	Attitudes		
	Opposed	Neutral	In favour
Nuclear station in UK	52.2	26.5	21.3
Nuclear stations in SW	58.0	26.0	15.5
Nuclear station locally	75.1	14.5	10.4
Coal-fired station	63.5	23.5	13.0
Windmills	37.6	22.9	39.5
Chemicals factory	90.3	7.0	2.7
Any industrial development	62.3	21.5	16.2

Note: Data from 'four counties' study, data set 2-V. $n = 290$

Evaluating technologies: how many alternatives?

An apparently straightforward question, but one with important implications, is whether people's evaluations of any single technology are influenced by the number of other technologies which they are asked to judge at the same time. In other words, are people's expressed attitudes towards nuclear power influenced by whether they are asked about nuclear power alone, or have also to say what they think about coal, oil, wind power or whatever? If people give systematically different answers, depending on the context in which they are asked a question, then it follows that these answers can only be interpreted meaningfully if one is told the context in which the questions are asked. This may seem an obvious point, but it is constantly overlooked in the reporting of opinion poll data, for instance. To be told that, say, 55 per cent of a sample approve of nuclear power means very little in absolute terms if different wordings of the same question yield different levels of approval. As we have seen, more or less specific questions about nuclear developments can produce different levels of acceptance, even though they intercorrelate. It may be easier to say that you approve of nuclear energy than to say that you approve of the building of more nuclear power stations, and easier to say that you approve of a new power station somewhere else than that you want one in your own neighbourhood.

The influence of question content or wording on questionnaire responses is reasonably well recognized. However, as we argued in Chapter 1, most opinion poll surveys of nuclear attitudes have restricted themselves to a limited number of simple questions. Because of this, there has been rather less systematic research on the effect of the context in which individual questions are presented (but see Schwarz *et al.*, 1987). One potentially important feature of context is simply the number of alternatives to be considered. We therefore conducted a set of experiments, by manipulating versions of the questionnaires distributed within our 'four counties' study, to investigate the influence of this variable. Our hypothesis was that judgment of the importance or desirability of any given energy technology would be higher, the fewer alternative technologies that are presented for judgment at the same time. Put differently, if people have to consider several alternative technologies alongside each other, they are likely to assign less importance to any one technology than if they consider that technology in isolation.

Before relating this prediction to more psychological research, it is worth noting the parallels here with much of the debate on nuclear energy and other environmental issues. Much of the case in favour of nuclear development runs along the following lines. We need electricity. There is a legitimate argument to be resolved over the priority to be attached to the need for new generating capacity as compared with the need for conservation and less wasteful use of existing supplies, but this is essentially a question of how much capacity is required, not whether we need such capacity at all. However it is produced, electricity does not come for free, either in terms of financial or environmental costs. Pointing out that there are costs associated with nuclear energy, therefore, is not in itself an argument for halting nuclear development. It has to be shown that the costs associated with any other technology capable of generating electricity nationally on the scale required are the same or less than those associated with nuclear power, and are likely to remain so for a considerable period of time. This (so it is argued) has not been shown. There are no alternatives to nuclear power that deliver the same benefits in relation to costs and can yield comparable amounts of power on a national scale. The antinuclear case (or one version of it) disputes the cost-benefit advantage of nuclear power and asserts that there are feasible alternatives, either singly or in combination. (Indeed, the very word 'alternative' is often used with a special meaning in relation to environmental issues, being identified with promotion of 'renewable' energy sources such as wind and wave power, and more generally with an avoidance of waste and a rejection of materialist consumerism. The implicit message here is again that there are alternative ways of doing things and these deserve to be taken seriously. To avoid confusion, though, we shall use the term 'alternative' simply as meaning 'other'.)

How might the influence of the number of response alternatives be interpreted from a more theoretical point of view? The literature of most relevance is that concerned with social judgments under conditions of uncertainty. Tversky and Kahneman (1974) have argued that people's processing of uncertain information is characterized by a reliance on various simplificatory strategies or rules of thumb, referred to as 'cognitive heuristics'. As mentioned briefly in Chapter 2 (in connection with residents' predictions of where a new power station would be built), one such heuristic, termed *anchoring and adjustment*, describes a tendency for incomplete adjustment of initial estimates. For example, Tversky and Kahneman report that subjects' estimates of the number of African states in the United Nations were influenced by first having to say whether the correct answer would be higher, lower, or the same as an arbitrary starting number. The higher the starting number, the higher were subjects' subsequent estimates. Presumably subjects assumed that the experimenters would not have asked for comparisons with a given standard if that standard was ridiculously out of range. By the same token, asking about a large number of alternatives might convey the message that all the alternatives mentioned deserve consideration, whereas asking about very few might imply that those not mentioned were less important. One possible consequence of such cues provided by question context could be that judgments of relative importance are 'anchored' to an inferred (rather than explicitly imposed) starting point, implying that all alternatives mentioned are roughly comparable in importance. Thus the more alternatives presented, the less absolute importance should be attached to any single one.

Another heuristic of potential relevance is that termed *availability*. According to Tversky and Kahneman, judgments of whether something is probable (and hence important) depend on the ease with which relevant thoughts can be accessed from memory or imagination. (More recent discussions tend to prefer the term *accessibility*, since not all information that is 'available' in memory is necessarily easily 'accessible'.) The more 'thinkable' something is, the more probable it will appear to be. Fischhoff *et al.* (1981) have discussed the possible biases that could be induced by this heuristic into people's judgments of risks associated with nuclear power and other hazardous processes and activities (see Chapter 6). Many factors may influence the accessibility from memory of particular thoughts or the ease with which alternative possibilities can be imagined. However, it is possible that explicit mention of a given alternative could increase its availability, whereas its exclusion from the context of judgment could make it less available, and more likely to be ignored. Distinguishing between these alternative interpretations would require a study in which individuals were led to think about a range of alternatives without using all of these as anchors or judgmental standards.

Context manipulations

The questionnaires distributed to members of the district samples (*A*, *B* and *C*) of the 'four counties' study included a section dealing with comparisons between different energy technologies (see Van der Pligt, Eiser and Spears, 1987a). The format of this section was varied (irrespective of locality) to generate three separate experimental designs. In all cases, this section comprised a (randomly ordered) list of different possible sources of energy, about which respondents were asked three questions: (a) how much of the United Kingdom's electricity supply was produced by each source; (b) how much they predicted would be so produced by the year 2000; and (c) how much they would ideally like to see produced from each source. All questions were answered in terms of percentage estimates. In each of the three experiments, the manipulation of interest was the number of sources of energy included in the list.

Experiment 1 involved a longitudinal design, including all respondents contributing to data set *1-A* who also responded at time 3 (data set *3-A*). Complete data for the relevant items was obtained from 184 respondents. At time 1, all respondents were presented with a list of five energy sources. These were oil, nuclear, hydro, coal and solar/wind/wave power. A sixth response category was labelled 'other' and the instructions were that the percentage estimates should add up to 100 across the six categories in response to each of the three questions. At time 3, respondents were allocated at random to one of five conditions, in each of which they were presented with just one of the original five energy sources and had to give three percentage estimates as before. (The introduction to the questionnaire explained that it included 'some questions which may seem similar to the ones you have seen before, but it is very important to know whether or not people's attitudes have changed'.)

Our prediction was that, irrespective of the type of estimate required and irrespective of the particular energy source in question, individuals would give higher estimates of the amount of energy produced (now, in the future or ideally) by a given source when that source was presented singly than when it was included in the context of other forms of energy. As can be seen from the mean percentage estimates in Table 4.6, this prediction was strongly supported

94

(overall, $p < 0.001$). The change scores were significantly ($p < 0.001$) greater for judgments of preference than for estimates of either present or future contributions (means were 28.2, 10.2, 14.7 respectively).

Table 4.6   Estimated and preferred prevalence of energy sources as a function of context (mean percentages): experiment 1

Energy source	Number of sources named		
	Five	One	Difference
Oil			
Estimation now	27.7	38.3	10.6
Estimation AD 2000	14.4	25.7	11.3
Preference	8.6	27.6	19.0
Nuclear			
Estimation now	19.0	30.4	11.4
Estimation AD 2000	39.0	54.8	15.8
Preference	12.8	32.2	19.4
Hydro			
Estimation now	0.9	23.4	12.5
Estimation AD 2000	14.9	36.7	21.8
Preference	19.9	64.8	44.9
Coal			
Estimation now	42.0	55.2	13.2
Estimation AD 2000	28.3	42.6	14.3
Preference	16.2	47.9	31.7
Solar/wind/wave power			
Estimation now	2.3	5.4	3.1
Estimation AD 2000	8.1	19.8	11.7
Preference	37.4	70.1	32.7

Experiment 2 also involved repeated testing. Analyses were based on 220 respondents from sample *B*, who provided complete data for the relevant items at time 2, and on a sub-sample of 134 who did so when followed up at time 3. At time 2, individuals were allocated at random to one of three conditions, in which they had to give estimates of the following energy sources: (a) nuclear; (b) nuclear, coal, plus 'other'; (c) nuclear, coal, gas, geothermal, hydro, oil, solar, wave and wind energy, plus 'other'. (Note that solar, wave and wind energy were distinguished as three separate categories, unlike in Experiment 1). At time 3, all respondents received, at random, one of the two versions they had not received at time 2. In other words, those originally presented with version (a) now received either version (b) or version (c), and so on. An important difference from Experiment 1 was thus that the size of the list was unconfounded with the time of presentation (whereas in Experiment 1, all the single item versions were presented after the longer list).

We first considered the data from the larger (time 2) sample in terms of a 'between subjects' design, that is, ignoring all time 3 responses. This design allows two kinds of comparison along the lines of those in the previous experiment. First, estimates of nuclear energy can be compared in the context (a) of nuclear alone, (b) of nuclear and coal, and (c) of nuclear and eight other named sources. Second, estimates of coal can be compared between conditions (b) and (c). Table 4.7 presents the means.

Table 4.7   Estimated and preferred prevalence of energy sources as a function of context (mean percentages): experiment 2

| | Number of sources named | | |
Energy source	(a) One	(b) Two	(c) Nine
Nuclear			
Estimation now	26.2	23.9	18.8
Estimation AD 2000	53.4	43.3	31.8
Preference	41.0	20.5	16.2
Coal			
Estimation now		56.4	39.1
Estimation AD 2000		37.4	25.8
Preference		40.7	15.4

As can be seen, estimates on all three scales were highest for nuclear in condition (a) and lowest in condition (c). Similarly, coal received higher estimates in condition b than in condition (c).

Included in this experiment were 134 individuals who provided ratings at both time 2 and time 3. If we just consider estimates of nuclear energy, since this was the only source rated by all respondents on both occasions, it is possible to split the sample into two groups. In one group, respondents were presented with a larger number of energy sources, together with nuclear energy, at time 3 than time 2, whereas the other group judged a reduced number of sources at time 3 compared with time 2. The results showed that the first group gave lower ratings at time 3 (mean changes being −0.9, −8.8 and −7.6 for present, future and ideal estimates), whereas the second group gave higher ratings (mean changes being respectively 9.7, 21.3 and 12.5).

Experiment 3 was conducted at time 3 on sample C, with 164 respondents providing complete data. Four different lists of energy sources were used: (a) nuclear; (b) coal; (c) nuclear, coal, hydro, oil plus 'other'; (d) nuclear, coal, hydro, oil, wave, wind plus 'other'. This design allows for comparisons of estimates of nuclear energy in contexts involving (a) one, (c) five or (d) eight categories (including 'other'). Similar comparisons can be made for coal across conditions b, c and d. Additionally, comparisons can be made between conditions c and d for hydro and oil. The means are shown in Table 4.8 and tell a similar story to that of the previous two experiments. In general, larger contexts yield smaller estimates of any given source. However, the differences between conditions c and d were not entirely consistent with this trend, with the effects of context on ratings of hydro and oil being nonsignificant. The effect of context on preference for nuclear energy was also nonsignificant, for the first time in any of these analyses.

Table 4.8   Estimated and preferred prevalence of energy sources as a function of context (mean percentages): experiment 3

Energy source	Number of sources named		
	(a/b) One	(c) Four	(d) Seven
Nuclear			
Estimation now	27.9	24.9	17.2
Estimation AD 2000	51.8	36.5	39.3
Preference	27.6	21.0	16.1
	($n = 44$)	($n = 47$)	($n = 33$)
Coal			
Estimation now	51.0	38.3	42.4
Estimation AD 2000	32.7	26.0	21.6
Preference	47.9	32.3	23.8
	($n = 40$)	($n = 47$)	($n = 33$)
Oil			
Estimation now		22.9	22.9
Estimation AD 2000		14.7	10.2
Preference		15.3	11.7
		($n = 47$)	($n = 33$)
Hydro			
Estimation now		8.9	14.9
Estimation AD 2000		15.0	12.8
Preference		22.9	16.8
		($n = 47$)	($n = 33$)

Interactions between context and attitude

All three of the experiments just described provided the opportunity to see if the effect of context on ratings was equivalent for individuals holding different attitudes. This question only applies to estimates of nuclear energy, since this was the only source with respect to which attitudes were measured. As in some of the analyses (on the village samples) reported in Chapter 3, individuals were split into three attitude groups on the basis of their responses to the single item requiring self-ratings of attitude to nuclear energy in terms of seven categories. Those describing themselves as 'very strongly opposed', 'strongly opposed' or 'opposed' were designated as anti, those describing themselves as 'neutral' comprised a neutral group, and those describing themselves as 'in favour', 'strongly in favour' or 'very strongly in favour' were designated as pro. This self-report measure correlated 0.83, 0.84 and 0.73 respectively in the three experiments with an alternative measure of attitude, the seven-item Likert scale described in Chapter 2. Responses (in terms of four categories from 'not at all' to 'very much') to the question 'How personally involved in this issue of a possible new nuclear power station are you?' indicated that the neutral groups saw themselves as least involved, with the antis being more involved than the pros.

Overall, the findings of the three experiments were that neutrals were most, and antis least influenced by the effect of context. Experiment 1 included only 43 individuals who rated nuclear energy on both occasions. Averaged over the three types of estimate, the mean changes from time 1 to time 3 were 17.6, 20.0 and 10.1 for the pro, neutral and anti groups respectively. Because of the small sample size, however, these differences were not statistically reliable. The mean changes for preference ratings were 25.4, 26.2 and 6.9 respectively (a marginally significant effect), implying that those opposed to a new nuclear power station were least affected in their preference ratings by the context manipulation.

In the second experiment, analyses of each of the three estimates showed significant interactions between attitude and the three context conditions (at $p < 0.005$ for present estimates, $p < 0.001$ for future and ideal estimates). As can be seen from the means presented in Table 4.9, the pro respondents appeared the least affected by context in their present and future estimates, but the most affected in their preference estimates. The neutral group was the most affected with respect to present and future estimates, whereas the minimal difference in preference shown by anti respondents as a function of context reflects a low preference rating across all three conditions.

In the third experiment, equivalent analyses were performed comparing conditions (a), (c) and (d) (one, five and eight categories). On the preference estimates, the means for the (a) minus (d) differences were 9.0, 21.7 and 7.8 for the pro, neutral and anti groups respectively, reflecting a significant interaction between attitude and context.

Table 4.9   Estimated and preferred prevalence of nuclear energy: the effects of context and attitude (mean percentages): experiment 2

Attitude	Number of sources named		
	(a/b) One	(b) Two	(c) Nine
Estimation now			
Anti	18.9	24.3	14.1
Neutral	38.0	23.7	18.7
Pro	26.3	23.3	26.6
Estimation AD 2000			
Anti	43.3	41.4	21.7
Neutral	69.6	42.6	23.9
Pro	54.3	48.7	50.2
Preference			
Anti	5.6	4.1	3.4
Neutral	56.8	28.7	21.3
Pro	74.2	49.6	31.7

The analyses of the other two estimates showed similar trends, but nonsignificantly. Separate analyses showed a significant influence of context within the neutral group for present estimates as well as for preferences.

Anchoring or accessibility?

The main finding from these experiments was that estimates of the proportion of the country's electricity supply contributed (now, in the near future, or ideally) by any given source of energy varied inversely with the number of energy sources specifically named and presented for judgment. In other words, the format of the questionnaire produced large differences in the ratings given both by different individuals and by the same individuals at different times. At very least, taking a purely methodological perspective, this sounds a cautionary note over too literal an interpretation of the absolute levels of approval of any technological development suggested by responses to opinion polls and surveys. Individuals may be confident that, looked at relatively, they would prefer, say wave power to nuclear energy. However, the translation of such relative preference into ratings on an absolute scale is a process that is extremely vulnerable to the influence of context.

The most plausible interpretation of such context effects seems to be that they reflect one or more of the information-processing strategies referred to as 'cognitive heuristics' by Tversky and Kahneman (1974). Much of the experimental literature on cognitive heuristics has been built upon demonstrations of fallible reasoning by student subjects presented with abstract or hypothetical problems (Nisbett and Ross, 1980). Although the biases found are often large and reliable, queries could be raised about their applicability in real-life contexts where individuals have more at stake than out-guessing an experimenter. Hogarth (1981) has argued that heuristics are relevant to decisions outside the laboratory, but in ways that are essentially adaptive for dealing with an uncertain environment. Our data show a quite consistent tendency for estimates to be influenced by the number of alternatives presented for judgment, in accordance with our prediction. The estimates concerned a real issue of personal relevance for our respondents. At the same time, it appears that those with less clear-cut opinions were most influenced by the context manipulations, whereas those with more definite viewpoints were influenced less. The estimates required were not of probabilities as such, but of the contributions, in percentage terms, of different energy technologies to the national electricity supply. Nonetheless, the judgments appear to reflect a decision strategy (or strategies) similar to some found in studies of probability and frequency estimation. In particular, there is strong encouragement from our data for the idea that respondents 'anchored' their estimates to an arbitrary starting point which involved all presented technologies being treated as more or less equal.

101

The phrase 'more or less' is not offered apologetically, since the contributions of the different technologies, now and in the medium term, are actually extremely unequal. The contribution of nuclear energy, around 16 per cent at the time of testing, tended to be overestimated. The contribution of renewable sources, while recognized as small, was also overestimated, although one could hardly expect otherwise: the likely national contribution of renewables, according to official estimates, was predicted to be only about 1 per cent by the year 2000. Conversely, the predominant position of coal at the time (roughly two-thirds of the total) was drastically underestimated. In other words, judgments of the present and future situations greatly underplayed the distinctions between the different technologies.

We cannot rule out the possibility that this is also partly a reflection of accessibility, since, as will be seen in Chapter 5, the potential value of renewable energy technologies was a matter of considerable (and favourable) media interest. It is also worth noting that a comparison of preferences with actual and future estimates pointed to a general wish for a decrease in reliance on nuclear energy and fossil fuels, but a substantial increase in the use of renewables. We did not set up the three experiments to distinguish between the relative contributions of, say, the heuristics of availability (accessibility) and of anchoring and adjustment. (In fact, there is a general question whether various heuristics, distinguished originally in terms of their effects, actually reflect distinct as opposed to common processes.) Even so, the anchoring and adjustment heuristic could account for most of our effects. The assumption required is that respondents took as their starting point or 'anchor' a working hypothesis that energy sources given equal mention were of broadly comparable importance, and certainly of greater importance than sources not specifically mentioned at all. In the next chapter, we shall be looking at the kinds of 'mention' afforded to different energy technologies by the media. A possible implication of our findings here could be that increased mention (media attention) produces a greater impression of importance, possibly over and above effects due to the persuasive content of the coverage itself.

In fact, subsequent research by Van Schie and Van der Pligt (1994) favours an interpretation of such effects in terms of the anchoring and adjustment heuristic rather than accessibility. They conducted a series of experiments in which subjects had to rate the causal contribution to acid rain of different categories of environmental pollution. The basic condition required subjects to give percentage estimates, adding to 100 per cent, for the categories Agriculture, Traffic, Industry, Private Households and Other Factors. Other conditions added to the number of specific causes presented by subdivision of one or more of theses categories (e.g. Agriculture was divided into Animal Farming, Crop Farming, and Glasshouse Horticulture). As in our experiments, adding more response items reduced the percentage assigned to any one item. Viewed the other way round, any given category received a higher total percentage if its subcategories were specified. However, this effect was essentially confined to conditions where subjects had to give separate ratings to these extra items (subcategories). Merely listing these extra items within an aggregate category without demanding separate estimates did not lead to an increase in the total percentage assigned to the category as a whole. It is difficult to see how the availability heuristic could account for such effects, since the listing of extra items should increase their accessibility from memory, regardless of the form of rating. In other words, the data imply that the ease of retrieval of alternatives from memory is far less important than whether such alternatives constitute response categories to be used in the judgment.

Conclusions

In the psychological literature on context effects, a general theme is that people use one item of experience as a comparison standard for judging other items (Eiser, 1990). Our experiments on the number of response alternatives were designed to get people to think about different sources of energy which might not immediately have come to mind as relevant to the estimation tasks which they had to perform. These findings need to be set against those of the Dorset studies reported earlier in the chapter, in which the comparisons were not between abstract policy alternatives, but real developments within the residents' own communities. We had mixed results in our attempt to test predictions from the work of Kahneman and Tversky (1979) and Fischhoff (1983) on 'decision framing'. Approval of new oil wells was somewhat dependent on whether residents considered that they would be having a new power station too; the majority of residents were more favourably disposed to oil than nuclear development. However, nuclear attitudes were less contingent on the prospects for oil development.

Rather clearer results emerged from our 'proximity' study, in which the sample was divided on the basis of how close residents lived either to the nuclear plant or oil wells. In the case of attitudes to a new nuclear power station, we found preferences for any new station to be built elsewhere in the region or the country, rather than locally. This preference was stronger among those living near to the nuclear site at Winfrith. In the case of attitude to oil wells, although drilling in the North Sea was preferred to inshore or onshore drilling, a local development was somewhat preferred to one further away on the south coast, especially among those living closest to the existing oil wells at Wytch Farm. Proximity of residence to a given development thus seems to lead to more definite evaluations (whether positive or negative), presumably attributable to its greater salience.

Analyses comparing the importance of specific anticipated impacts of each development further emphasized the relative optimism about oil as compared with nuclear development. However, such local concern as there was about new oil wells showed many parallels with concern about a new nuclear power station, albeit in a more muted form. Just as we have argued that much local opposition to nuclear expansion is not specifically antinuclear in its motivation, so a lack of enthusiasm for more oil drilling does not depend greatly on disadvantages specific to oil exploration and extraction. People's feelings about the visible and natural environment in which they live and about the character of their community can lead to the witholding of support for any development seen as likely to change these cherished goods for the worse. Concerns about major disasters and dangers to health will be expressed along with more local concerns, but for those people who might tolerate similar developments elsewhere rather than near their own homes, they do not appear to be the most important criteria. The data that lead us to this conclusion, however, were obtained in settings where such disasters and health risks may have been viewed as rather conjectural, particularly in the case of oil.

The question of how attitudes to nuclear power are changed by news of accidents and evidence of risk is one to which we shall return in Chapter 6. But first we need to broaden our discussion in a different direction. We have described how different attitude groups view the prospects of nuclear and other developments, but we have not explicitly considered how these different groups view each other. We have related individual differences in attitude to the differential salience of aspects of proposed developments, but have not considered the importance attached to these aspects by the media in their coverage of energy issues. These are the themes to be addressed in the next chapter.

Chapter 5

Representations
of the Nuclear Debate

In the preceding chapters we have taken a more or less direct look at
the nuclear debate in terms of distributions and determinants of
attitudes to nuclear power, from the wider international level down to
the local context of our own research focus. In Chapters 3 and 4 we
also started to examine some of the more social psychological
dimensions of this debate such as how attitudes can determine the
perceived salience of different aspects of the issues in ways that
reflect and reinforce these attitudes. In the present chapter we
continue this focus by exploring how the represention of the issues
surrounding nuclear power, both in the media and by the public
themselves, can further help to explain the polarized nature of this
debate. Such representations add a further important social
psychological dimension because they inform and move the debate as
well as simply describing it. The purpose here is also to flesh out in
more detail the structure and content of the nuclear controversy. This
is more than simply the distribution of 'pro' or 'con' views toward
the 'attitude object' of nuclear power in some homogeneous
population. It comprises quite a rich context of adversarial groups
and organizations, agendas and ideologies, representing the various
positions and interests involved. Gaining a better profile of the two
sides and how they are represented, how they characterize nuclear
power, and also how they view each other, is of some importance if
we are to acquire a more complete understanding of the nuclear
debate from a social psychological perspective.

In the first section of this chapter then, we examine media
coverage of nuclear power. The aim here is twofold. First of all, the
media form an obvious and important source of the information and
views that help to shape the public attitudes and opinion which form
the central theme of this book. We consider briefly the nature of this

107

relationship with particular reference to the impact of the media on attitudes for an issue as controversial as nuclear power. More generally, the media can be seen as providing a window through which to view the nuclear debate itself, providing an important perspective on the dimensions, actors and organizations that comprise it. We concentrate specifically on local newspaper reporting of nuclear energy and its alternatives as examined in our own research. This focus allows for some sensitivity to regional variations in the impact which nuclear power makes and how it is regarded, particularly in relation to the local policy and planning developments (see Chapter 2). In the second section of this chapter we shift our focus from the media as such, and return to examine the importance of attitudes in how people construct and represent nuclear issues, and particularly the two sides of the debate. In Chapter 3 we examined the structure of people's attitudes, and the differences between attitude subgroups in their views, while in Chapter 4 the focus was on the comparison of attitudes to nuclear power with those to other developments. Here we consider how attitudes to nuclear power may shape social perceptions and attributions in ways likely to lead to a further polarization and perseverence of these attitudes.

The mass media and public opinion in the nuclear debate

Attitudes and beliefs concerning an issue such as nuclear power do not exist in a vacuum, but are influenced by a range of sources in the social context, such as peers, significant others, local groups and so on. However, for an issue of the technical complexity of nuclear energy, the media are likely to be a particularly rich and influential source of information that may not be directly available from these other agencies. In short, the media may form both a springboard for and a measure of the debate itself. We begin with a brief and general consideration of the two-way relationship between media and public opinion in the context of nuclear power before proceeding to a more detailed analysis of (local) media coverage in its own terms.

The effects of the mass media on public opinion have long been debated (see e.g., Lemert, 1981; Lowry and DeFleur, 1983; MacKuen and Coombs, 1981; McGuire, 1986; McQuail, 1987;

Roberts and Maccoby, 1985). Early research suggested that the media had more or less direct influence on its audience, seeing it as a relatively passive recipient of the media message. However this rather pessimistic model of the unquestioning 'dupe' public has subsequently come in for much criticism on both empirical and theoretical grounds. A degree of consensus has since emerged that the effects of the media are more limited or at least mediated by a host of other demographic, social and even attitudinal factors so that it is difficult to isolate them as directly causing opinions. The media may reflect cultural norms and values as well as defining them (e.g., Paletz, Reichart and McIntyre, 1971; Breed, 1958; Murdock, 1974). MacKuen and Coombs (1981) conclude that the media tend to reinforce and confirm people's existing views and Lemert (1981) argues that the effects of overt media persuasion on attitude change are usually small or negligible. On the other hand, Lowry and DeFleur (1983) maintain that effects whilst mediated can still be powerful especially in terms of the more cumulative influences often missed by many typical 'one-shot' experimental studies. Although a particular item may not have a significant impact, media coverage which chronically presents certain ideas or images can also 'cultivate' a particular climate of opinion (cf. Gerbner and Gross, 1976; Gerbner et al., 1980). Moreover failure to demonstrate an impact on individual attitudes or behaviour does not necessarily mean that the media have no effect at all; general social perceptions of the world may be more sensitive to media coverage than belied by 'personal' views or judgments (e.g., Tyler and Cook, 1984).

The persuasion literature in social psychology suggests a whole range of source, message and perceiver characteristics that are likely to mediate influence effects (e.g., Petty and Cacioppo, 1986; Zanna, Olson and Herman, 1987), and a complete analysis of mass media influence should perhaps be considered in this framework. It is beyond the aim and scope of this chapter to explore such issues in any detail, but clearly this question is a very complex one and attempts to abstract the media from their cultural context and accord them a direct causal role are bound to be fraught with difficulty. Our main objective here is to use an analysis of media coverage as an indicator of the nuclear debate itself, rather than regarding it as a pre-formed stimulus variable which causes public opinion. The media may also be a source of the views of agencies or interests that lie beyond the bounds of what might loosely termed public opinion, but which may also play a powerful role in this debate. Nevertheless we start by discussing briefly a few factors which suggest the public's perceptions of nuclear issues and their representations in the media are likely to be particularly closely related, especially at the local level.

While media effects might not be overt or automatic in the sense of changing attitudes directly, the actual weight and focus of coverage may have an effect on perceptions of the issue in 'setting the agenda', by directing public attention to the issue and framing its terms of reference (McCombs and Shaw, 1972; McLeod, Beckers and Byrnes, 1974). In short, while not telling us what to think, the media can certainly determine what is made salient, and thus channel the nuclear debate. 'Newsworthy' events are not picked up by the media any more automatically than their coverage is assimilated by the public. In this sense media coverage involves 'manufacturing' news or at least 'gate-keeping' in so far as editors and journalists decide when to make an item into an 'event' or an 'issue' for public consumption. Sometimes environmental hazards or risks may be known about for years both by experts and communities before the media decide to 'break' the story (Peltu, 1985). Peltu suggest that 'whistle swallowing' may be just as important as 'whistle blowing' in newsmaking, and suppressing or simply not reporting events can often be related to proprietal and advertising interests of the media or even to governmental influence. This raises the question of the effects of actual media bias; although the audience may not always be influenced by what is said they clearly cannot be influenced by what is not said. The status attached to the specific medium may also accord a certain status or importance to the message for the target, affecting the impact of the message.

Agenda-setting and the degree of media attention may be of especial importance for public perceptions of nuclear power. Even if coverage does not change opinions directly, making the public aware of the issue may provoke greater thought and discussion on the issue, both of which *have* been shown to polarize attitudes (e.g., Tesser, 1978; Turner *et al.*, 1987). Furthermore there is evidence that greater coverage of nuclear issues can tip the balance in the antinuclear direction because the public may be made more aware of the potentially catastrophic consequences associated with nuclear technology (e.g., Mazur, 1984). Indeed, research on risk-perception suggests that the media may play an important role in leading people to overestimate the probability of certain risks by disproportionate coverage, and by the attention attracted to them. Such risks may be more 'available' in the media, but partly because of this also more cognitively accessible, leading people to overestimate the prevalence of nuclear hazards (Fischhoff *et al.*, 1981; Lichtensten *et al.*, 1978; Slovic, Fischhoff and Lichtenstein, 1982; see also Chapter 6). Peltu (1985) also argues that the particular language, symbols and imagery used by journalists to communicate and popularize nuclear issues may inevitably lead to negative evaluations and reactions without being overtly biased. For such reasons, Mazur (1984) has maintained that heightened nuclear coverage of nuclear power, even where balanced or slightly pronuclear, will tend to increase public fears and thus opposition.

Another indirect way that the media may influence public opinion is via what Katz and his colleagues refer to as the *two-step flow of communication* (Katz, 1957; Katz and Lazarsfeld, 1955). According to this theory, certain figures in the community serve as opinion leaders, mediating influences between media and public. This channel of influence seems particularly relevant to the local context of opposition to nuclear developments. As we have seen in the earlier chapters such plans witness the emergence of many *ad hoc* local opposition groups. In this context, the activists involved are bound to become recognizable local figures and may function as opinion leaders for sections of the community by interpreting and communicating developments in the local media. The flow of information may be mediated in other subtle ways also. Whilst disputing the thesis of direct attitude change, Lemert and his colleagues argue that the media can facilitate 'mobilizing information', that is 'information which allows people to act on the attitudes which they might already have' (Lemert *et al.*, 1977, p. 721). Again this notion is highly relevant to the local context and especially the issue of planning and building new nuclear power stations. Thus while it is known that attitudes on this issue tend to be highly polarized (see also Chapters 1,3 and 4) and resistant to

111

attempts at influence, the media may potentiate or thwart activity in the community by revealing mobilizing information. Again the local media are likely to be best placed for imparting this information because the national press and TV are unlikely to be as detailed in their coverage of such locally relevant issues. Mobilizing information is particularly important in the context of nuclear power, because this is an issue over which people and groups become active (see e.g. Chapter 1). Mazur (1984) and Peltu (1985) have also suggested that activists and interest groups may speak through and use the media as well as simply responding to it. This does not just take the form of letters to the editor. Peltu (1985) describes how the media may be used and to some extent manipulated by activists on both sides of the nuclear debate, via press releases and briefings, 'stunts', and so forth. Over 90 per cent of UK environmental organizations had received press coverage in 1979/1980 (Lowe and Morrison, 1984).

Finally, public attitudes are above all sensitive to media coverage of 'continuity-breaking' events such as Three Mile Island, and Chernobyl as well as other less dramatic controversies (e.g., Mazur, 1984; Nealy, Melber and Rankin, 1983; MacKuen and Coombs, 1981; see Chapters 1 and 6). Clearly it would be misleading to put down such effects exclusively to media influence because such event-related shifts in public opinion are likely to have been substantial, regardless of the medium of communication. Nevertheless, it is important not to underestimate the role of the media coverage in heightening public awareness of such issues. Again, the quality and quantity of coverage is not necessarily assured and may depend on a host of other and even 'chance' factors. For example Peltu (1985) suggests that the extremely high-profile coverage of the Three Mile Island accident (40 per cent of networked television news-time in the week of the accident) was at least partly due to the accessibility of the area to reporters and journalists nearby. The coincidental release of the film *The China Syndrome* also stimulated the public imagination at this time, making the issue even more 'hot'. In sum, such events can dramatically increase the coverage accorded to nuclear power, and as we have seen, by their very nature (accidents, cancer scares, etc.) they are far more than likely to decrease public confidence in nuclear technology than the reverse. The very criteria for newsworthiness such as drama and negativity (cf. Peltu, 1985) are

112

likely to count against nuclear technology confirming Mazur's assertion that no news is good news as far as the defenders of the nuclear industry are concerned.

To summarize then, while the influence of the mass media on the public may not be direct or immediate there are nevertheless a range of mechanisms by which the media can have considerable impact on its audience, especially with respect to controversial issues such as nuclear power. In presenting our own research in this field the aim is not to demonstrate a causal or even reciprocal link between media representations of the issue and public opinion, although the foregoing analysis does suggest ways in which this is likely to occur, and that the two may be particularly closely geared at the local level. Our more modest objective is to examine the structure and evaluative tone of media representations of nuclear power, and the agencies and organizations responsible for these evaluations. This should then provide a broader but also more detailed picture of the nuclear debate at a time (and place) relevant to the rest of the research reported in this volume.

Local newspaper coverage of nuclear energy and its alternatives

As suggested above the effects of media on public attitudes are likely to be especially relevant at the local level, and this level of analysis also permits comparisons across region in relation to nuclear siting policy. We therefore conducted a content analysis of local daily newspaper coverage across the UK (Spears, Van der Pligt and Eiser, 1986a; 1987), using a sampling frame of 103 newspapers. Our sample comprised all significant coverage of nuclear power issues (i.e., more than one inch column space) for the first six months of 1981. We decided that it would be important to include comparison with another energy technology because understanding and evaluation of any particular energy option only really becomes practically meaningful in relation to some possible alternative option that could (at least potentially) fulfil similar needs (Farhar-Pilgrim and Freudenberg, 1984; see also Chapter 4). More generally the aim was to provide a control or contrast category against which to interpret representations of nuclear technology. Given that nuclear power is often presented as a virtually boundless source of energy to be contrasted with the limited and depleting fossil fuels of coal or oil, we chose instead alternative and renewable energy sources (e.g., wind, wave, solar power) as an appropriate comparison category. Although, very different in many ways, nuclear power and 'renewable alternatives' could both be seen as possible candidates for solving our long-term energy needs.

The general aim of our content analysis was to examine the content and structure of evaluations regarding these two technologies, as well as the sources of such evaluative appraisals. Clearly these technologies can be evaluated according to a range of different criteria. Thomas *et al.* (1980) performed a factor analysis of belief items concerned with energy technology and found five underlying dimensions which were used to structure our content analysis:

1. Economic aspects
2. Environmental aspects
3. Technological aspects
4. Future/political aspects (indirect risk)
5. Physical/psychological aspects (direct life risk)

In addition to the source of the article we recorded the evaluative tone of headlines, and also the evaluative structure of the main text of the articles according to the scheme outlined above. However we split the economic dimension into short- and long-term costs or benefits, because a single dimension might have produced a tendency to undervalue costs incurred for long-term gain ('investment', for example), or to overvalue false economies. The coding scheme was finalized after extensive piloting in order to attain satisfactory intercoder reliability. A miscellaneous category for evaluations which did not fit into these dimensions or which were not specified was also added, resulting in a seven-category coding scheme.

The sentence formed the unit of analysis for the actual text of articles. Each sentence was coded for positive or negative evaluations of the two technologies and was assigned to one of the seven categories in the scheme. Neutral, indeterminate and merely descriptive statements were ignored. If the source of these evaluations could be inferred from the article this was also recorded. Sources of evaluative appraisals were classified according to a ten-category scheme as follows:

1. Pronuclear industries and organizations (e.g., CEGB, BNFL)
2. UK central government
3. Advisory institutions and commissions
4. UK local government
5. Independent institutions (e.g., universities, non-governmental research institutes, 'experts')
6. Media: 'active' press and TV
7. Independent inquiries
8. Public (e.g., personal/public opinion, letters to the editor)
9. Antinuclear pressure groups and organizations
10. Pro-alternative pressure groups and organizations

A distinction was also made between articles appearing simply to report brute fact, and more polemical or evaluatively oriented value-based articles. The development stage of the technology was coded if this was mentioned or could be inferred. Attention variables (headline size, article length and the number of pictures, if any) were also recorded. All categories attained very high degrees of inter-rater reliability. These, together with more precise details of coding procedures and examples are to be found in Spears *et al.* (1986; 1987).

Description of the sample. For the 6 month monitoring period 939 articles were detected from 89 of the 103 newspapers; 488 (52 per cent) contained evaluations concerning nuclear power, 352 (37.5 per cent) contained evaluations concerning alternatives and 99 (10.5 per cent) contained evaluations concerning both. The *East Anglian Daily Times* accounted for 14 per cent of articles, more than double the output of the next most prolific source: the *Western Morning News* (6.9 per cent), followed in turn by the *Evening Star* (Ipswich) (6.6 per cent), the *North Western Evening Mail* (3.5 per cent), the *Dorset Evening Echo* and the *Morning Telegraph* (both 3.3 per cent).

An important distinction from the point of view of our research was whether newspaper coverage corresponded to places that were or

115

were not affected by present or impending nuclear developments, and the actual or potential planning of a new nuclear station in particular. We refer to these throughout as the *affected* and *unaffected* samples respectively. The affected sample comprised coverage from the *East Anglian Daily Times*, the *Western Morning News*, the *Evening Star* (Ipswich), and the *Dorset Evening Echo*. The two East Anglian papers qualified for this sample because of the CEGB's then proposal to build the UK's first PWR at Sizewell in Suffolk (since approved by a Public Inquiry). The two papers in the South-West qualify on grounds of the CEGBs plans to site a new nuclear power station somewhere in this region, as investigated in our own research (see e.g. Chapter 2). Coverage from these sources comprised quite a reasonable sample ($n = 289$ or 31 per cent of the overall sample) to compare with coverage from the remaining articles (i.e., the unaffected sample).

It is instructive simply to compare the breakdown of coverage of nuclear and alternative issues across these samples. In the affected sample 66.4 per cent of articles were concerned with nuclear power compared with 17.3 per cent and 16.3 per cent for alternatives and both together, respectively. The analogous figures for the the unaffected sample were 45.5, 46.5, and 8.0 per cent respectively, so that in contrast with the affected sample, articles concerned with alternatives were in a majority here ($p < 0.001$).

Evaluations of nuclear and alternative technology. Perhaps the most salient source of evaluation is to be found in headlines. Figure 5.1 presents the percentage distribution of positive and negative headlines for nuclear power and alternatives (excluding evaluatively neutral headlines) for affected and unaffected samples. In both samples it is clear that negative evaluations of nuclear power substantially exceed positive evaluations although the reverse is the case for alternatives. This difference in distribution for nuclear versus alternatives for the sample as a whole was very significant ($p < 0.001$) and this pattern is even more accentuated in the affected sample. Turning to the main textual content of articles, Figure 5.2 depicts the percentage of articles in both subsamples containing *any* evaluative coverage of nuclear or alternative technologies.

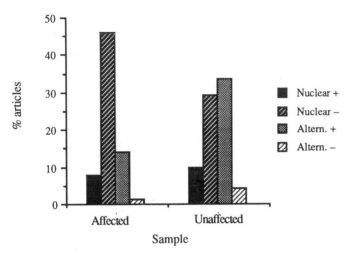

Figure 5.1   Percentage of articles with positive and negative headlines about nuclear and alternative technology as a function of sample.

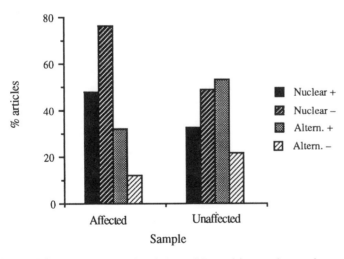

Figure 5.2 Percentage of articles with positive and negative statements about nuclear and alternative technology as a function of sample.

The pattern is identical to that for headlines, although the differences are less accentuated. However, this index is also rather conservative because it does not give any indication of the overall evaluative tone of articles, nor the distribution of evaluation across the seven dimensions. To take these factors into account we examined the actual weight of positive and evaluative coverage for nuclear and alternatives across the seven content dimensions, for both affected and unaffected samples. Table 5.1 summarizes these data and specifically presents the percentage of articles in the respective samples containing net negative and net positive evaluations (a net positive evaluation is given by a preponderance of positive over negative statements within a given article and vice versa for a net negative evaluation. See Spears *et al.*, 1986 for more detail regarding these data, and also for information concerning the net *extremity* of evaluations).

In terms of the economic dimension both nuclear power and alternative technology tend to fare more badly in the short than the long-term, but alternatives are also evaluated more positively than nuclear power on these dimensions. Over 20 per cent of articles in the affected sample are evaluated negatively with respect to nuclear power on the short-term economic dimension, compared to less than 7 per cent for alternatives, although this difference tends to even out for the unaffected sample (14.5 and 12.8 per cent respectively). Hardly any articles result in a net positive evaluation on this dimension. Net positive evaluations emerge more clearly on the long-term economic dimension. However, while the percentage of positive articles regarding nuclear power (10.4 and 8.0 per cent of the affected and unaffected samples respectively) tends to be balanced out somewhat by negatively slanted articles (9.7 and 3.8 per cent respectively), a far greater proportion of alternative-related articles are positively weighted on this dimension (i.e., 17 and 34.6 per cent for the affected and unaffected samples respectively) compared to a negligible proportion of 'negative' articles.

Turning to the dimension of environmental risk, this is clearly a dimension of greater relevance to nuclear power than alternative technology where there are very few net evaluative appraisals. Nuclear power is, however, overwhelmingly negatively evaluated in terms of this dimension with 31.1 per cent of articles in the affected

118

sample and 18.6 per cent of articles in the unaffected sample resulting in net negative evaluations compared with 5.9 and 4 per cent respectively that are on balance positive. A similar, though less extreme relationship holds for the future/political risk dimension. Again, this is more or less of no relevance to alternative technology, but produces a preponderance of negative articles (15.9 and 11.1 per cent of the affected and unaffected sample respectively) compared to positive articles (3.1 and 0.3 per cent) about nuclear power.

Table 5.1. The percentage of articles containing net positive and net negative evaluations as a function of evaluative dimension and sample

Dimension	Sample	Nuclear		Alternative	
		Positive	Negative	Positive	Negative
Short-term	Affected	1.0	20.4	1.7	6.9
economic	Unaffected	2.3	14.5	1.8	12.8
Long-term	Affected	10.4	9.7	17.0	1.0
economic	Unaffected	8.0	3.8	34.6	2.6
Environ-	Affected	5.9	31.1	1.7	1.0
mental	Unaffected	4.0	18.6	4.0	1.7
Future/	Affected	3.1	15.9	1.0	0.0
Political risk	Unaffected	0.3	11.1	2.0	0.0
Technology	Affected	5.2	10.4	20.1	0.3
	Unaffected	5.8	7.8	44.3	0.6
Direct	Affected	8.0	28.4	1.0	0.0
life risk	Unaffected	4.9	20.9	1.9	0.6
Unqualified	Affected	16.3	52.6	16.6	3.1
	Unaffected	12.3	19.1	19.7	3.5
Overall	Affected	15.6	65.1	29.1	2.8
	Unaffected	12.8	35.5	48.8	4.0

The 'technology' dimension is in contrast of greater relevance to alternative energy sources than to nuclear power. Over 44 per cent of articles in the unaffected sample and over 20 per cent of articles in the affected sample were positive about alternative technology on this dimension while the proportion of negative evaluations is negligible. However, for nuclear power, the percentage of negative evaluations outweighs the percentage of positive ones in both samples (10.4 vs. 5.2 per cent, and 7.8 vs. 5.8 per cent for the affected and unaffected samples respectively). The direct-life risk dimension is irrelevant to evaluation of alternative technology, but again results in a surfeit of net negative evaluations of nuclear power (28.4 and 20.9 per cent of articles in the affected and unaffected samples respectively) and relatively few positive ones (8 and 4.9 per cent). Finally the miscellaneous or unqualified dimension tends to highlight the general trend emerging over the content-specific dimensions (compare with the summarized evaluations in Table 5.1) with the greater preponderance of articles evaluating nuclear power negatively, but alternative technology positively.

Although this pattern of evaluative coverage is quite clear-cut, we made a more direct comparison of the two technologies by analysing those articles which covered both technologies ($n = 99$). Consistent with the data from Table 5.1, six of the seven dimensions resulted in net negative evaluations of nuclear power for this subsample whereas an equal number of dimensions were net positive for alternatives. In comparing the two technologies, alternative energy sources are evaluated significantly more positively on the short-term economic, long-term economic, future/politicial risk, technological and unqualified dimensions and this difference is marginally significant on a sixth (environmental risk). Direct life risk is the only dimension not to result in a significant difference favouring alternatives. When comparing the evaluative coverage in the affected and unaffected subsamples, the overall mean negative evaluation of nuclear power is significantly greater in the affected sample (-3.22 compared to -1.49, $p < 0.01$). The greatest contributors to this difference are the economic and unqualified dimensions. Alternatives, although evaluated net positively, are also evaluated significantly less positively in the affected than the unaffected sample (0.97 compared to 2.41, $p < 0.01$). This discrepancy is accounted for most by the economic and technological dimensions (see Spears et al., 1986, for more details of the differences on particular dimensions).

To summarize the overall pattern of evaluative coverage, a fairly clear-cut picture emerges. In general terms, nuclear power tends to be evaluated negatively and alternative technology positively. With the caveat that alternatives do not seem to feature on the evaluative dimensions relating to risks, this pattern recurs fairly consistently across the different dimensions used to characterize energy issues. Finally, coverage in the affected sample is relatively more negative than in the unaffected sample for both energy sources, although again, in absolute terms nuclear power is evaluated far more negatively.

Sources of evaluations. Looking purely at the evaluations of nuclear power and renewable alternatives still leaves out of the frame what or who is responsible for such appraisals. This question was examined by considering the percentage of articles in each subsample (affected/unaffected) where particular sources of negative and/or positive evaluations are indicated (n.b. while the percentage values may appear generally small, they are expressed as a function of the *overall* subsample and not just as a proportion of purely 'nuclear' or 'alternative' articles). Due to the number of source categories and evaluative dimensions, further multiplied by evaluation and sample, these results are quite detailed so we shall only attempt to pick out the salient features of the data here. A complete account can be found in Spears *et al.* (1987).

For nuclear power, the pronuclear industries and organizations category is consistently the most prolific source of positive appraisals. For example, in the affected sample this source is responsible for positive evaluations on the short-term economic dimension in 4.5 per cent of articles, 8.3 per cent of articles for the long-term economic, 9.0 per cent for environmental risk, 2.1 per cent for future/political risk, 4.8 per cent for technological risk, 12.5 per cent for direct life risk, and 20.4 per cent of articles for unqualified positive appraisals. This percentage drops somewhat in the unaffected sample where, as one would expect, there are fewer articles about nuclear power in any case. However, the important comparison point is that no other single source even approaches these percentages in terms of their support for nuclear power, in either sample. Excluding the unqualified category, the highest figure reached by another source is 2 per cent and the vast majority of source/evaluative dimension combinations actually result in percentages less than 1 per cent or do not register at all. Nevertheless it is interesting to note that the UK central government is the only other source besides the nuclear industry itself to be more often responsible for positive than negative appraisals.

The distribution of sources of negative evaluations of nuclear power provide a sharp contrast with those for positive appraisals. First of all the 'negative' sources are much more widely distributed than for positive appraisals. The 'public' in particular is a very substantial detractor, especially in the affected sample. This source was responsible for miscellaneous negative evaluations of nuclear power in over 30 per cent of articles in this subsample and between 4 and 14 per cent of articles as regards the specific dimensions of evaluation. The environmental risk and direct life risk dimensions figured particularly prominently in this respect (13.8 and 13.1 per cent of articles respectively). Again this general profile, though similar, is far less pronounced for the unaffected sample. Another noteworthy point of comparison is that the pronuclear source tends to produce a higher percentage of negative evaluations of nuclear power in the unaffected than the affected sample. Moreover, compared with the negative appraisals of nuclear power, the pronuclear and UK government sources tend to be *less* frequent sources of negative appraisals than positive ones, whereas the converse is true for all other categories of source.

The distribution of evaluative sources for alternative energy technology provides a sharp contrast to the pattern for nuclear power, as might be expected from the different overall pattern of evaluation described earlier. Unlike the positive appraisals of nuclear power, the sources responsible are fairly evenly spread and not just located within the clearly partisan category (in this case the pro-alternatives groups and organizations). In fact in the unaffected sample, the 'independent institutions' appears to be the most prolific source; over 17 per cent of articles in this sample received acclaim from this quarter on the technological dimension, with 10.9 and 9.5 per cent of articles in this sample also receiving support from this category on the long-term economic and unqualified dimensions respectively. Nevertheless, in the affected sample, the pro-alternatives source is the one most consistently responsible for positive appraisals, where once more the long-term economic, technological and miscellaneous categories seem to be the main dimensions of acclaim (see Spears *et al.*, 1987).

122

Sources of negative statements concerning alternatives are few and far between, reflecting the paucity of such evaluations generally, as described above. Nevertheless, the most important sources here are 'independent institutions' and 'pro-alternatives', especially in the unaffected sample. The highest percentage for a particular source/dimension combination was for pro-alternatives on the short-term economic dimension in this subsample (3.7 per cent of articles) but the norm was well below 1 per cent per cell.

To summarize, the distribution of sources of evaluations produces a fairly high degree of consistency across the seven dimensions. The pronuclear industries and organizations are by far the most prolific source of positive appraisals of nuclear power whereas the public exceed even the antinuclear movement *sui generis* as the most frequent detractors of this technology. By contrast with positive appraisals of nuclear power, sources of negative appraisals of nuclear power and also positive appraisals of alternative technology are much more broadly based, and not confined to what might be regarded *a priori* as partisan groups. For example, 'independent institutions' figure relatively prominently as sources in these two cases. Few sources figure in terms of negative appraisals of alternative technology, reflecting the general lack of such evaluations *per se*.

The development stage of the technology. Analysis of the development stages of the two technologies to which articles refer reveal characteristic differences. Articles concerned with nuclear power concentrated mainly on the policy decision phase (prior to a public inquiry) and on nuclear power when fully operational and feeding the national grid. Comparing across subsamples the policy phase is predictably far more relevant to the affected sample (over 60 per cent of articles), than the unaffected sample (fewer than 10 per cent of articles). Articles about alternative energy sources tended to be concerned with piloting and promoting the technology (e.g., reports of exhibitions and open-days) as well as its operational use.

Attention variables. Attention variables such as headline size and the length of articles are important because they are likely to mediate the impact and influence of particular coverage. Headlines in the affected sample were significantly bigger, averaging 5.35 column widths compared with 4.43 for the unaffected sample ($p < 0.001$). There was no significant effect of technology nor any significant interaction. In terms of length of coverage, articles in the affected sample were also significantly longer on average (13.6 inches of column space per article compared with 11.5 for the unaffected sample). Also, articles concerning alternative technology were significantly longer than articles about nuclear power (13.9 compared to 10.9).

Type of coverage: factual or polemical? Although the fact-value distinction is a difficult one to make in the nuclear debate, it may nevertheless be informative about the sort of coverage the two technologies attract. We classified articles on the basis of whether they made declarative statements of fact alone, or whether they introduced polemic, debate or any points of subjective value (pre-testing produced a high degree of intercoder reliability). Comparing across technology, 83.2 per cent of purely nuclear articles fall into the 'polemical' category compared to 53.6 per cent of articles concerned purely with alternative technology ($p < 0.0001$). Comparing across subsamples, 86.8 per cent of articles from the affected sample contain polemic compared to 67.8 per cent for the unaffected sample ($p < 0.0001$). Thus there are more value-based or polemical articles about nuclear power than about alternative energy sources and more in the affected than in the unaffected sample.

Summary and conclusions
It is perhaps not surprising that the geographical spread of coverage on these energy issues and concerning nuclear power in particular, is most concentrated in those localities with possible sites for a possible new nuclear power station. Both the then ongoing debate surrounding Sizewell, together with the CEGB's interest in a site in the South-West explored in our own research, made these areas natural foci of local media interest. The question then concerns the evaluative structure and content of such coverage.

The overwhelming finding is that nuclear power is evaluated negatively, especially in comparison with alternative renewable energy sources. This holds relatively consistently across the seven evaluative dimensions used to characterize energy issues, although the risk dimensions seem particularly critical for the evaluation of nuclear power, and equally irrelevant to the evaluation of alternative technology. Predictably perhaps, coverage of nuclear power tended to be even more negative in the affected sample, and this is consistent with the finding that people are more opposed to building a nuclear power station in their locality than they are to nuclear power in general (see e.g., Chapter 2). The difficulties of inferring cause-effect relations notwithstanding, this correspondence gives further credence to the close relationship between local media coverage of nuclear power and local opinion as discussed earlier.

The finding that nuclear power tends to attract less factual and more value-based coverage tends further to reflect the general controversy surrounding the nuclear issue. Moreover, because such controversy is bound to be heightened in areas confronted with new nuclear developments, it is consistent that polemical coverage should be highest in the affected sample. This supports Frankena's (1983) claim that controversy provokes a shift from fact-based to value-based issues in energy policy. The finding that articles in the affected sample had larger headlines than those in the rest of the sample also serves as an indicator of the prominence of such issues for those who are directly affected by them.

The polemical and polarized nature of the debate is also illustrated by considering the distribution of sources of positive and negative appraisals of the two technologies. The nuclear industry and also central government are the only consistent advocates of nuclear power (it is perhaps interesting that the advisory institutions and commissions authorized by this government, do not share their overall positive evaluation of nuclear power, at least according to the local press coverage). By contrast, opposition to nuclear power is spread across a number of different categories suggesting a broader consensus antagonistic to nuclear power than in favour. The general public exceed even the antinuclear movement as the greatest detractors of nuclear power, and this contribution is even more marked in the affected sample, consistent with our more direct attitude survey research (see Chapter 2). On the other hand, support for alternatives is quite broad-based in comparison to nuclear power, proliferating in the 'independent institutions' and 'pro-alternatives' categories in particular.

125

To summarize then, evidence from local press coverage provides clear evidence of polarization in the nuclear debate, both in terms of the spread of evaluations and the structure and quality of coverage. Nevertheless this polarization does not seem to be evenly balanced, a fact made clear by the contrast with alternative energy sources. Coverage is heavily skewed in the antinuclear direction, and not simply in terms of the sheer weight of negative appraisals, but also in terms of the broader spectrum of groups and organizations that comprise this opposition in comparison to the pronuclear lobby. This overall pattern would seem to add further empirical weight to Peltu's (1985) assertion that no news is good news for the nuclear industry. Having acquired a more detailed picture of the structure and content of the nuclear debate through the eyes of the local media, we now return to consider more directly people's own attitudes and interests on this issue, and how these influence their perception of it.

Social perception and judgment of participants in the nuclear debate

A clear theme that emerges in the analysis of the media coverage is the heavily polarized and evaluative nature of the nuclear debate. As the issue becomes more controversial and heated, it is likely to see a shift, not just from neutral to evaluative coverage, but from facts to values *per se* (cf. Eiser and Van der Pligt, 1979; Frankena, 1983). This implies that the perception of the actors involved itself becomes more partisan but also informed by particular knowledge, interests and opinions. This also concurs with evidence in Chapter 3 that people with opposing attitudes towards nuclear power tend to see different aspects of the issue as salient or important. In the remainder of this chapter we explore further ways in which attitudes and involvement structure and influence people's perceptions of the policy-making process concerning nuclear power, and the different factions involved. Once more we concentrate on illustrative findings from our own research rather than attempt a comprehensive review of the social psychological literature. Specifically we first consider the influence of how one's attitudes and interests can influence attributions for policy decisions in the local context of our own research. Then we proceed to examine how attitudes, involvement, knowledge and expectations can influence one's perception of the distribution of attitudes more generally.

Attributions for nuclear decisions: Attitudes and interests
In the context of our survey research (see Chapter 2), we were interested in how local residents might perceive the decision-making process, and specifically the CEGB's choice of site, both from the perspective of their own attitude and in terms of how they were personally affected by the decision (see Eiser, Van der Pligt and Spears, 1989). These data were collected in the 'four counties study' in September and October, 1982, shortly after the final announcement of the proposal to build a new nuclear power station at Hinkley Point in Somerset, with the option of further developments at Winfrith in Dorset left open. These survey data comprise both 'district' and 'village' samples (i.e. samples 3-*A*, 3-*B*, 3-*C*, 3-*V*, 3-*W*, and 3-*X*; see Chapter 2). Specifically, we measured respondents' perceptions of the CEGB's attention to a series of aspects and the perceived influence of a range of agents on this decision, as a function of their general attitude to nuclear power and how their community was affected by the decision. Aspects of the decision-making process, combined with our design, resulted in four categories of community:

1. A *control* group made up of residents in South Devon. This community had no existing nuclear power station and was not threatened with one in the future.
2. The *eliminated* group. This comprised residents in communities originally nominated as potential future sites by the CEGB, but subsequently eliminated, and included all the Cornwall samples and those in Dorset ruled out by the decision.
3. The *delayed* group. This sample included the remainder of the sample in Dorset, namely those living in the locality of Winfrith for which the CEGB was keeping its options open.
4. The *chosen* group. This sample included those living close to Hinkley Point, the site definitely chosen by the CEGB.

127

As we have already seen in Chapter 3, attitude was systematically related to community — that is, the eliminated group were more strongly opposed to nuclear power (in terms of their general attitudes) compared with the other three samples ($p < 0.001$). For the following analyses however, respondents were grouped into pro- and antinuclear attitude categories by means of a median split on the overall sample.

Perceived reasons for the CEGB's choice of site. First we were concerned with how respondents viewed the reasons for the CEGB's decision (choice of site), in terms of attention to a range of factors (e.g., national economic need, improved regional electricity supply, local opposition/acceptance; see Table 5.2 below). Respondents as a whole differentiated very significantly between different reasons ($p < 0.001$), attributing most attention to technical feasibility and operating costs and least to local opposition/acceptance and local economic impact. However, more interesting for our present purposes is to see how the perceived priorities of the CEGB were seen to vary as a function of attitude and sample. From Table 5.2 it is clear that attitude (pro vs. anti) has a significant effect on attributions of perceived attention by the CEGB on all but one of the eight aspects (namely local opposition/acceptance). On all other dimensions pronuclear respondents rated the CEGB's attention higher than did antinuclear respondents. Turning to the effect of locality, the respondents in the chosen community gave higher ratings to construction and operating costs, but lower ratings to regional supply considerations.

Perceived influence of different sources on the CEGB's decision. Estimations of the perceived influence of different sources revealed that central government was perceived as the most influential (see Table 5.3 for a full list of the sources presented to respondents). Pros attributed less influence to more antinuclear sources (the local press, environmentalists and local action groups). Residents of the eliminated communities gave the highest and those of the chosen community the lowest, ratings to local action groups, local newspapers, and the environmentalist movement. The eliminated group also rated central government as less influential.

From these illustrative data it is clear that there is no homogeneous consensus about the CEGB's priorities, or the perceived influences on them, but that these are to some extent mediated by attitudes and local involvement or vested interests. The effect of attitude suggests a general 'evaluative consistency response set' such that the more pronuclear the residents' attitudes, the more credit they gave the CEGB for attending to a whole range of aspects (with the exception of local opposition/acceptance). Evidence that the more a source supported the person's attitude, the more influential it was perceived as being also fits with this evaluative consistency principle. Attitudes may thus have a motivational input into how people perceive the actors and influences in the debate, in ways which support, reinforce or reflect them.

Table 5.2 Attributed attention of CEGB to different aspects in choice of site as a function of attitude

| | Attitude | | |
Aspect	Anti	Pro	F (1,899)
National economic need	4.18	5.37	65.67**
Improved regional electricity supply	4.00	5.37	56.92**
Local opposition/acceptance	3.51	3.67	1.22
Environmental impact	3.47	4.08	18.24**
Construction/operating costs	5.19	5.56	7.47*
Local economic impact	3.24	3.79	14.68**
Improved national electricity supply	4.84	5.60	31.55**
Technical feasibility	5.52	5.85	6.94*
n	437	471	

Note: Scale ranges from 'no attention at all' (1) to 'very much attention' (7)
* $p < 0.01$, ** $p < 0.001$.

Turning to the variations between the localities, differences between the eliminated group and the delayed and chosen communities were particularly evident in terms of perceived influences on the CEGB's choice of site. Thus, in the 'reprieved' communities, local action groups and the environmentalist movement tended to be seen as more influential and the central government less so as regards the decision. These differences may be partly attributable to the generally more negative attitudes in this group, although significant differences for sources where attitude produced no or weaker effects (e.g., central government) suggest that this cannot be the whole story. A number of possible explanations for this pattern are possible. One concerns logical inference in terms of the congruence between actions and outcomes (Brewer, 1977). That is, local action groups were judged as more influential by those communities whose activities could be seen as successful in achieving their intended goals. It could be that respondents in this eliminated sample generalized from perceptions of their own success and overestimated the influence of these sources on the decision in general. Cognitive accessibility or 'availability' (Tversky and Kahneman, 1973) could also be employed to help explain this effect (see also Chapter 6). Whereas the eliminated communities were involved in a relatively high level of protest activity leading up to the decision, the chosen community (i.e. Hinkley Point) was not even originally on the CEGB's list, so that for this community there was no real cause for protest prior to the decision. Differential recall of the activity of environmentalists and local action groups may therefore simply reflect the cognitive accessibility of their actual activity in the locality. In addition to these cognitive or information-processing explanations there is also a third more motivational possibility. Given the general local opposition towards building a nuclear installation in one's vicinity (see Chapters 1 and 2), residents had strong vested interests in the decisions they were asked to explain. To the extent that particular sources were, or could be seen as being, more successful in protecting these interests at a local level, these may have been ascribed greater influence.

To summarize then, it is possible to discern in these data a number of motivational and cognitive factors at play which influence perceptions and judgments of a crucial policy decision, both in terms of residents' attitudes, and as a result of the involvement or vested interests arising from how the decision affected one's own community. In the following section we take a more systematic look at the effects of attitude and involvement on perceptions of the nuclear debate, and specifically how these factors influence the perceived distribution of opinions in this context.

Table 5.3 Perceived influence of different sources on the CEGB's choice of site

Aspect	Attitude		F (1, 1896)
	Anti	Pro	
Local government	3.83	3.82	0.00
National newspapers	2.70	2.48	3.58
Local action groups	3.87	3.33	12.62**
Central government	5.09	5.02	0.22
Environmentalist movement	3.94	3.60	5.47**
Local newspapers	3.06	2.71	6.11**
Independent scientists	4.10	4.14	0.07
n	444	460	

Aspect	Locality				F (3,896)
	Control	Eliminated	Delayed	Chosen	
Local government	3.79	3.69	3.78	4.02	1.40
National newspapers	2.77	2.56	2.79	2.23	5.39***
Local action groups	3.73	4.41	3.76	2.51	44.39***
Central government	5.33	4.27	5.07	5.52	22.55***
Environmentalist movement	4.00	4.22	3.93	2.93	22.36***
Local newspapers	2.99	3.09	3.05	2.41	7.45***
Independent scientists	4.13	3.92	4.29	4.12	1.39
n	94	382	185	243	

Note: Scale ranges from 'no influence at all' (1) to 'very great influence (7)
* $p < 0.05$, ** $p < 0.001$.

Attitudes, involvement and perceived consensus

If attitudes and involvement in this issue can influence perceptions of particular policy procedures and outcomes, the question is whether such factors colour, or indeed inform, perception of the nuclear debate more generally. In this section we consider ways in which one's attitudes and beliefs can influence perceptions of others attitudes and opinions. This notion draws on a burgeoning literature within social psychology that suggests that social perception and information processing is influenced by our own attitudes, 'schemas' or expectations , for a range of cognitive and motivational reasons (e.g., Fiske and Taylor, 1991; Nisbett and Ross, 1980).

One such phenomenon of direct relevance to the present issue is the so-called *false consensus effect* (Ross, Greene and House, 1977), namely the tendency to overestimate the proportion of other people in some reference category or population who share one's own attitudes or opinions (the typical contrast category being estimates of those people subscribing to the opposite view). Indeed, in some of our own work we have consistently demonstrated this effect in the context of attitudes to nuclear power and related issues (e.g., Spears and Manstead, 1990; Spears, Eiser and Van der Pligt, 1989; Van der Pligt, Ester and Van der Linden, 1983). There are a number of possible mechanisms responsible for such an effect. It may be caused by the salience or accessibility of one's own viewpoint, leading the person to over-generalize this to the population as whole. Selective exposure to similar others who share your view may be another important determinant which is perhaps of particular relevance in the small fairly close-knit communities of our own research. However this tendency may be also have a motivational foundation — seeing greater support for your own view may be reassuring especially if this view reflects strong vested interests (cf. Crano, 1983), as it invariably does for this particular issue. The purpose is not to provide a test of such explanations here, and the effect is probably over-determined in any case (see Marks and Miller, 1987; Mullen *et al.*, 1985; Mullen and Hu, 1987; Spears and Manstead, 1990, for recent reviews of this literature). The point is simply to illustrate that one's attitudes can influence the perception of the nuclear issues in quite fundamental ways. This is of particular importance for an issue such as nuclear power where local and public opinion are not just 'dependent variables' but crucial inputs into the debate that may even be taken into account in policy-making.

The false consensus effect could thus help to explain and reinforce polarization in the nuclear debate by helping to justify the cause (or entrench the views) of people on both sides. Such polarization may itself be perceived by the participants, especially by those who are more involved in the issue. In addition to overestimating the prevalence of one's own position, Judd and Johnson (1981, 1984) have suggested that people with strong attitudes in particular will tend to overestimate the prevalence of *both* attitude extremes. Again, this would tend to reflect and encourage polarized *perceptions* of consensus. Shortly we shall return to consider this issue in relation to the additional factors of involvement and knowledge. First however, we briefly explore how one's own attitude can influence the perception of consensus across *different* groups.

In a series of experimental studies, we were interested in how one would perceive the distribution of attitude positions in different groups as a function of one's own attitude or opinion regarding nuclear power (Spears, Van der Pligt and Eiser, 1985; 1986). In a typical study we presented participants with a series of slides depicting the opinions to nuclear power ostensibly expressed by the residents of two towns in South-West England. These towns were described as being confronted with the potential building of a new nuclear power station in their locality (our subjects were not drawn from any of the communities involved in our research, but were students at Exeter University) and were labelled as Town A and Town B in the slides so that actual knowledge or expectations about them would not bias participants' perceptions. In fact the views depicted were pre-tested and experimentally manipulated and were not drawn from residents of towns at all. However, given the CEGB's plans, this provided a plausible cover story in the context of our research in which to explore the effects of attitude on such perceptions.

The basic mechanism explored in this research was a perceptual bias known as the *illusory correlation effect* (Chapman, 1967). Illusory correlation simply refers to the erroneous perception of a relationship between two (or more) dimensions. In the present case we were interested to see whether participants in our experiments would misperceive the distribution of attitudes within the two towns. A classic study on the illusory correlation effect in social psychology by Hamilton and Gifford (1976) showed that people tended to overestimate the incidence of infrequently occurring behaviours in a small and infrequently occurring group, when in fact there was no association between the group membership and this behavioural dimension within the stimuli presented. For example, when undesirable behaviours were in a minority, subjects tended to overestimate the proportion of these emanating from a small and infrequently occurring group and rated this group more negatively (Hamilton and Gifford used this phenomenon to help explain the stereotyping of minority groups). In terms of the underlying mechanism they suggest that because minority behaviours in the minority group form the rarest or most unusual combination, it is likely to be the most distinctive and accessible in memory (Tversky and Kahneman, 1973), resulting in inflated prevalence estimates upon recall. We were interested in developing this idea and seeing if illusory correlations could result from factors other than distinctiveness based purely on numerical infrequency.

In keeping with the preceding discussion of the false consensus effect, we hypothesized that people would perceive attitude positions congruent with their own as particularly salient and memorable, for both the cognitive and motivational reasons outlined above — especially for people with more involved attitudes. In terms of Hamilton and Gifford's illusory correlation paradigm then, we predicted that pronuclear subjects would tend to over-associate pronuclear attitudes with an infrequently occurring town, whereas antinuclear subjects would over-associate antinuclear positions with this town. Following Hamilton and Gifford we reasoned that this combination (salience due to self-relevance plus distinctiveness due to infrequency) would be the most distinctive and memorable during recall, thereby influencing people's prevalence estimates.

In one of these experiments we manipulated the frequency of statements such that a minority emanated from Town B and a majority from Town A, while an equal *proportion* of pro- and antinuclear statements came from both towns (Spears *et al.*, 1986). Results confirm our general prediction, namely that subjects tended to overestimate the prevalence of attitude positions congruent with their own attitude in the less frequently occurring town. Moreover this effect increased as a function of involvement or attitude extremity as predicted. Figure 5.3 presents data concerning two indices of illusory correlation as a function of attitude extremity (in this case illusory correlation reflects the degree to which perceivers overestimate the association between attitudes congruent with their own and the smaller, more distinctive, town. This association is summarized in a transformed *phi* coefficient).

These data illustrate that attitude and involvement 'operationalized as attitude extremity' can influence or bias the perceived distribution of attitudes across different groups, as well as within a single group or target population as for the false consensus effect. This has a number of implications. For example, it suggest that people in a minority group whose own viewpoint is also in the minority are likely to see their own views as particularly characteristic of their group (Spears, Van der Pligt and Eiser, 1985). This can function to increase a sense of ingroup cohesiveness and solidarity, a process that might be quite self-serving in confirming such people in their views. If realism tells you that your views are not shared by a large section of the population at large, you may still be able to console yourself with the belief that your own group or community think as you do. Moreover, if your group or community is confronted with the prospect of a new nuclear power station, this may be the relevant reference group in any case. The illusory correlation mechanism described here suggests there may be a cognitive as well as a more strategic or motivational foundation to such a phenomenon. There may also be a more general tendency to associate your own views with your own group, irrespective of infrequency (Spears *et al.*, 1986; Spears and Manstead, 1990) which would serve a similar 'bolstering' function.

Such a function may well be served by another class or determinant of 'illusory correlations', namely those based on prior expectations or schemata (e.g., Chapman, 1967). Research in this tradition shows that a prior expectation concerning a relationship between the two dimensions can bias subsequent perceptions so as to confirm the expected relation. Again this has been applied to the realm of social perception and stereotyping (e.g., Hamilton and

Rose, 1980), but the same principle could apply to the perception of group attitudes in the nuclear debate.

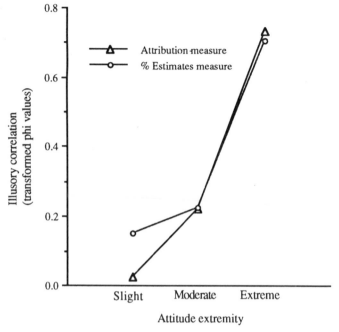

Figure 5.3 Degree of illusory correlation as a function of attitude extremity.

The opportunity to test this idea arose as a result of one of our experiments where it seemed that describing the more frequently occurring town (Town A) as larger, and the less frequently occurring town (Town B) as smaller, was interfering with the predicted illusory correlation effect described previously (Spears *et al.*, 1986, Experiment 1). It occurred to us that the pattern of results obtained might be explicable if subjects had an expectation that residents of small towns would be more antinuclear and residents of large towns less so, at least in the context of siting a new nuclear power station in their vicinity. We tested this idea in another experiment by presenting subjects with an *equal* number of pro- and antinuclear statements distributed equally across the two towns, but where we stated that Town A was *bigger* than Town B (Spears, Eiser and Van der Pligt, 1987). As predicted, an illusory correlation resulted in line with the expectation that residents of the smaller town (B) would be more antinuclear (see Table 5.4). Again these findings suggest a mechanism whereby people's own views or opinions may persevere, even in the face of apparently contradictory evidence.

Table 5.4. Percentage estimates of pro- and antinuclear residents in Town A and Town B.

Town	Antinuclear		Pronuclear	
	Estimated %	Actual %	Estimated %	Actual %
Town A (large)	44.19	50.00	55.81	50.00
Town B (small)	60.89	50.00	39.11	50.00

To summarize then, evidence from false consensus and illusory correlation paradigms suggest two important psychological mechanisms whereby people's own attitudes may influence social perception and specifically the perceived distribution of others' views in the nuclear debate. Moreover, the evidence suggests that this is likely to be even more pronounced to the extent that such attitudes are strong, extreme or involved. In other words, the nuclear debate is not only likely to be characterized by a polarization of views, but to the extent that this is true, this is likely to contribute to more divergent *perceptions* of consensus for both cognitive and motivational reasons. As we have argued, such perceptions may be quite functional for the individuals or groups involved. Such perceptions are likely to bolster opinions for both individuals and groups, as well as creating new expectations or beliefs that can themselves influence subsequent perception and judgment.

Nevertheless we should be wary about taking such arguments too far. Our data are illustrative of certain judgmental biases that may be at work in the nuclear debate, but these are unlikely to be the only factors involved. This line of reasoning also tends to characterize the person with strong and partisan attitudes as necessarily having a biased perspective on the issues. Although the preceding paragraphs might appear to lead in this direction, there is also good reason to suppose that those who are involved in the nuclear debate are likely to be more knowledgeable and in tune with such issues. This implies that they should be perhaps be *more* aware of the distribution of attitudes to nuclear power than neutral or uninvolved people. Another study concerned with consensus judgments addressed this question (Spears, Eiser and Van der Pligt, 1989).

Like the false consensus research, this study was concerned with people's 'self-generated' estimates of attitude distribution (rather than recalling perceived distributions as in the illusory correlation studies). However, it was also designed to evaluated the 'polarization prevalence' hypothesis of Judd and Johnson (1981, 1984) briefly described earlier. This idea extends the false consensus principle in suggesting that the more extreme or involved one's own attitude, the more likely one is to overestimate support for *both* attitude extremes. According to Judd and Johnson, this occurs because for someone with a strong attitude on the subject, attitude positions at both extremes are likely to be invested with 'affective intensity', rendering them more accessible for recall when it comes to generating prevalence estimates. However, in a range of experiments, we found very mixed support for this heuristic principle (Spears *et al.*, 1989).

Subjects were categorized by attitude strength and the prediction derived from Judd and Johnson (1981) was for an increase in estimates of positions congruent and incongruent with one's own attitude as attitude strength increases. However, our data suggested that while estimates of the pronuclear categories tend to increase as a function of attitude strength, estimates of the antinuclear position tend to decrease commensurately (i.e., for both pro- and antinuclear respondents). This pattern tends to argue both against the polarization effect proposed by Judd and Johnson, as well as predictions premised on false consensus principles. However this pattern could be explained in terms of differential knowledge associated with attitude intensity. Involved people (anti- as well as pronuclear) could simply be more aware that there is a significant proportion of people

138

sympathetic to nuclear power, and this knowledge may be better reflected in their estimates than uninvolved people. In other words, far from being more biased, more involved people from both camps may be more knowledgeable or aware of actual consensus distributions. In sum, although people who are especially involved in the nuclear issue may be prone to information processing and motivational biases, paradoxically it seems they can also be more informed and thus accurate in their consensus estimates (see Spears *et al.*, 1989 and Spears and Manstead, 1990 for a fuller discussion of this issue). How these apparently contradictory effects of attitude and involvement interact with each other is an interesting topic for future research.

To summarize then, a range of illustrative data from our research demonstrate the influence of attitudes, involvement and vested interests on the person's perception of nuclear issues, and particularly the perceived distribution of attitudes within and between groups. Although the last study described shows that we should be very cautious about characterizing those with involved attitudes as being 'biased', our results do point to a number of cognitive and motivational processes which may result in polarized perceptions, bolstering the polarized opinions on which they are based.

Conclusions

This chapter has been in two parts. First we examined how issues relating to nuclear power were depicted in the media, and specifically the local press. We then went on to consider how one's attitudes towards this technology affected perceptions of the issue, and especially the distribution of other opinions. Although seemingly different spheres, a common thread is how the nuclear debate is represented; at a cultural or 'macro' level in terms of the news media, and at the psychological or 'micro' level in terms of individual attitudes and perceptions. At both levels, evidence was found of considerable polarization on this issue. While it is true that nuclear power was overwhelmingly negatively evaluated in the local press, those few sources which were in favour, namely central government and the nuclear industry, are of course precisely those most in control

139

of policy-making and the technology itself. In short, polarization in this issue reflects more than simply a 'bimodal' distribution of attitudes 'for' and 'against' nuclear power in some homogeneous population; it also reflects structural conflicts between different institutions, factions and interests. Nevertheless, in considering this broader canvass, it is important not to lose sight of individual agents and their perspective on the issue, because collectively they make up the various constituencies and organizations concerned. Illustrative data from our own research shows how attitudes and involvement can colour but also inform social perceptions in various ways. Both of these 'macro' and 'micro' perspectives suggest a number of levels on which the inherent oppositions in the nuclear debate may be produced, represented and reinforced.

Chapter 6

Nuclear Accidents

Accidents and risks

Of all the psychological issues relating to nuclear power, the one that has received most widespread attention is that of how people interpret the likelihood of a nuclear accident and its consequences. If we can talk of the industry and the public holding divergent perceptions, perhaps nowhere is this divergence more evident than in their respective attitudes to nuclear risk. As we have stressed in previous chapters, attitudes concerning local environmental impact are firmly grounded in people's experience of their local environment, in expectations of how a particular piece of visual landscape will be altered, of changes in the character of a particular village, and many similar specific hopes and fears. For many people, however, there are also more general fears about serious accidents or pollution hazards. These may be based typically on what they have learned about nuclear safety from sources other than there own direct personal experience, as well as possibly on broader beliefs about industrial safety in general. If accidents occur at nuclear plants elsewhere, can one be sure that a local power station would be as safe as the industry claims? It is quite reasonable for such doubts to be raised, even if technical evidence can offer equally reasonable reassurance.

As we shall see in Chapter 7, 'expert' definitions of risk are formulated explicitly in terms of estimated probabilities of particular kinds of consequences. Although these consequences can be catastrophic, the estimated probabilities are generally so vanishingly small that the overall risk (consequence severity multiplied by probability) is regarded as negligible. According to the 'official' line, therefore, public apprehensions are attributable to an exaggerated perception of these probabilities. The research we shall describe in this chapter, however, supports a rather different interpretation. What matters is not whether ordinary people make inaccurate or accurate numerical estimates of remote probabilities, but how they try to make sense of the uncertainties with which they are confronted. One way they try to do so may be by questioning whether something that has actually happened could happen again, with possibly even worse consequences. Much of the debate depends less on strict probabilities than on the viability of drawing generalizations from one kind of accident or experience to another.

As we saw in Chapter 3, concepts such as 'psychological risk' and 'personal peace of mind' are important predictors of overall attitude. This suggests that feeling safe is a prerequisite for approval of nuclear power. In Chapter 1 we described national surveys, the results of which suggest that confidence in the nuclear industry can be damaged by news of accidents in other countries, as well as by anxieties concerning nuclear risks of a more military nature. All in all, it would appear that generalization, and the association of different kinds of information together with one another, are fundamental processes in the formation of attitudes.

As we saw in Chapter 5, news reporting is rarely favourable to nuclear power. Ordinary people will therefore 'know' a certain amount that is bad about nuclear power. Nuclear accidents make news, which is to say that it is 'known', at a general level, that nuclear accidents can occur. The issue is whether such general knowledge will be brought into play when assessing specific risks associated with different kinds of nuclear installations, including developments in one's own locality. Do people think that, if nuclear power is dangerous anywhere, it is dangerous everywhere? Or do they look for reassurance that their personal safety has not been compromised? It is for such reasons that we shall now look at evidence of how news of nuclear accidents and pollution can influence people's attitudes in different ways.

Lessons from Three Mile Island

The accident at the Three Mile Island (TMI) nuclear power station near Harrisburg, Pennsylvania, occurred in the early hours of 28 March, 1979. Following overheating of the reactor core, a large amount of coolant water was contaminated and remained on the floor of the reactor; radioactive krypton gas was trapped in the containment building and some radioactivity was released into the atmosphere. News reports for the first few days were incomplete, alarming and confusing. As a consequence, 60 per cent of those living within five miles of the plant chose to evacuate their homes. There followed a two week emergency period, including school closures and a ban on evening meetings.

Superficially, life returned to normal for most residents after this two week period. The actual amount of radioactivity released into the atmosphere (as distinct from the containment building) was calculated to be small and unlikely to lead to any excess cancer deaths (Flynn, 1979). However, the psychological impact was far more substantial. Stress levels among the local population were closely monitored with the use of both self-reports and physiological measures. The perceived threat and level of concern about emissions was typically rated as very high (Flynn, 1979, 1981). However, the information provided to the public, for instance about emergency procedures, was typically seen as inadequate (Kraybill, 1979). Such critical attitudes were held more strongly by younger and better educated residents. Feelings of demoralization were evident in the immediate aftermath but, according to Dohrenwend *et al.* (1979), had dissipated five months later.

Although the damaged reactor was shut down, occasional leaks of radioactive gas occurred over the following year. This provided an objective basis for continued feelings of stress and from uncertainty among residents about whether they had personally been exposed to radiation or not. Davidson, Baum and Collins (1982) observed deficits in performance on cognitive tasks reminiscent of symptoms of *learned helplessness* (Seligman, 1975). Physiological stress indicators and the reporting of physical symptoms were higher among local residents than in comparison communities 17 months later (Baum, Fleming and Singer, 1982; Collins, Baum and Singer, 1982) and such differences were still detectable 28 months after the accident (Davidson, Baum, Fleming and Gisriel, 1986).

In Chapter 1, we mentioned the impact of the TMI accident on attitudes expressed in national opinion polls. Consistent with the national picture, local attitudes to nuclear power became much more negative in the immediate aftermath of the accident, before falling back towards their previous levels some months later. However, a small but consistent negative change persisted (Rankin *et al.*, 1981). While national surveys in the US indicated less general confidence in nuclear safety (Kasperson *et al.*, 1980), levels of trust in the industry and in public officials were especially damaged among those living near to the plant itself (Dohrenwend *et al.*, 1979).

Taken as a whole, these findings suggest two lessons for our own research. The first is that the psychological impact of nuclear accidents and incidents can be far more serious than the proven consequences, assessed in terms of excess deaths from cancer and similar statistics. Fears of what might have happened, and uncertainty about actual details of what did happen, contribute particularly to this negative impact. These doubts relate to what is often termed 'psychological risk' (see Chapter 3). The second lesson is that general measures of attitude will show shifts in response to such events, but the absolute size of such shifts may be modest in absolute terms. Part of the reason may be that such events can often be interpreted in more that one way, so that individuals may persevere with their initial point of view even in the face of evidence which many others might regard as contradictory. As Lindell and Perry (1990) have remarked about the TMI accident:

'Opponents cited it as support for their belief in the inevitability of a catastrophic accident, while proponents found confirmation for their belief that defence-in-depth had made nuclear power plants safe enough for continued operation.'

Pollution from Sellafield

As a single event, the TMI accident may have been a turning point in public beliefs about nuclear safety. Its impact seems less related to the severity of the actual consequences than to what many people felt could have happened if luck had been less kind. Nonetheless, it generated a great deal of unfavourable publicity for the nuclear industry worldwide. One might not normally expect such attention to be given to a series of minor incidents, even if their cumulative public health effects were greater. Even an apparent cumulative danger can be newsworthy, however, if the relevant evidence is discovered or communicated in a sudden and dramatic fashion.

The nuclear reprocessing plant at Sellafield (formerly Windscale) in North-West England has not had an enviable history. There was a major fire on the site in 1957, followed by a sequence of minor incidents over several years. In November 1983, a television documentary *(Windscale — The Nuclear Laundry)* was broadcast nationally. The central allegation was that abnormal levels of cancer in the communities around Sellafield were attributable to radioactive discharges from Sellafield, and specifically that the incidence of childhood leukaemia in Seascale (the nearest village) was ten times the national average. The programme claimed that dangerous quantities of plutonium and other radioactive material had been discovered in houses in the neighbourhood of the plant. It also traced the paths taken by radioactive waste discharged into the Irish Sea to destinations as far away as Greenland. British Nuclear Fuels Ltd. (BNFL), the operators of the plant, challenged the evidence and denied any causal connection between the excess deaths and pollution from the plant.

Subsequently a government-commissioned inquiry (Black, 1984) confirmed that the cluster of excess deaths from leukaemia in the neighbourhood was sufficient to raise concerns about safety at Sellafield, although the statistical difficulties in drawing precise causal inferences from such data are considerable (Craft, Openshaw and Birch, 1984). The issue has remained controversial, with various explanations being suggested, including speculations about risks associated with population mobility and new housing developments. One recent hypothesis is that the excess leukaemia cluster might be attributable to occupational exposure rather than pollution of the environment outside the plant. If the children's fathers worked at the plant, their sperm could have been damaged by radiation (Gardner *et al.*, 1990).

The documentary itself, however, did not anticipate a debate of this kind, but rather presented a convincing picture of Sellafield as a major polluter of the surrounding coast and countryside, of BNFL as less than completely honest and open about the safety record of the plant, and of Seascale and nearby villages as communities blighted by personal tragedies. The programme attracted considerable publicity, even before it was broadcast. Expectations were raised that particularly shocking evidence would be revealed and BNFL prepared itself to defend itself against an extremely damaging challenge to its public image.

How might all this have influenced public attitudes to nuclear power generally among people living in other parts of the country? To answer this question we conducted a further postal survey among residents of communities in South-West England. Our aims were, firstly, to look at effects on residents' nuclear attitudes as a function of their knowledge about the programme; and, secondly, to look at residents' beliefs about the relevance of the evidence from Sellafield to the issue of the safety of nuclear installations nearer their own homes. In other words, the television broadcast provided us with an opportunity to look at how highly publicized information about nuclear power became integrated into ordinary people's knowledge-base and at how that knowledge-base was used to draw generalized inferences about nuclear safety in other contexts.

There was a further twist to the Sellafield story. Between November 10 and 14, 1983, there was a large unplanned discharge of radioactive material into the sea from Sellafield, resulting from the washing out of storage tanks. High levels of radioactivity were discovered in silt around the Sellafield pipeline by Greenpeace divers. Around November 22, radioactive waste drifted back to shore, contaminating a 200m stretch of beach. Radioactivity was also detected in clumps of seaweed brought in by the tide. The Department of the Environment then warned the public to keep away from beaches in the area until they had been decontaminated. As far as our study was concerned, we were too quick off the mark to catch much of any impact of this later incident (despite the widespread publicity it too received). Our questionnaires were distributed immediately after the broadcast in the first half of November, and 90 per cent of those returned were completed before the closure of the beaches.

146

The 'Nuclear Laundry' Study

The sample for this study (Van der Pligt and Eiser, 1988) consisted of residents of communities in South-West England who had responded, some 13 to 14 months previously, to the third phase of the 'four counties' study described in Chapters 2 and 3. As in our previous studies, the procedure involved mailing questionnaires together with a free reply envelope, and following up non-respondents about two weeks later. Of the 1085 names selected, 98 were no longer at the same address, but 811 (82 per cent) of the remainder responded. Six who omitted personal information were excluded, leaving a sample of 433 men and 472 women, with an average age of 47.8 years, whose responses could be matched with their answers given in 1982.

The questionnaire first asked respondents if they had seen the programme *Windscale —The Nuclear Laundry*. Responses were in terms of the categories: 'I saw all/most of it', 'part of it', 'no, but I heard about it through the news', and 'no, not at all'. A series of items then dealt with beliefs concerning the Sellafield plant and the extent to which it was responsible for the effects described in the programme. Specifically, respondents were asked if they believed the Sellafield plant had caused cases of leukaemia and other cancers, whether the incidence of cancers was higher around Sellafield than in the rest of the country, and whether the plant had caused cases of leukaemia and other cancers (as suggested) in Western Scotland and the Hebrides. Responses were in terms of five categories ranging from 'yes, definitely' to 'definitely not'. The next set of items asked for ratings of the extent to which cases of cancer around Sellafield had been caused by natural background radiation, routine discharge of radioactive waste into the sea, accidents at the plant, and other unknown factors. Respondents were then asked if they thought that the Sellafield plant had discharged plutonium into the sea, and whether it caused radioactive pollution of the local environment; they were then asked if they thought that equivalent effects were caused by the power station at Hinkley Point.

More general measures of attitude were then repeated from the questionnaires administered the previous year. These consisted of a single-item measure ('How would you describe your own attitude

147

towards nuclear energy?'; responses from 1 'very strongly opposed' to 7 'very strongly in favour'), and the eight general attitude statements about nuclear power used in earlier phases of the 'four counties' study. These were combined to form a single-attitude index with good internal reliability (0.86). A more specific measure of support or opposition (seven response categories) with respect to the building of a new nuclear power station in the South-West was also included.

We first of all looked to see if respondents had, on average, changed their attitude towards nuclear power since they were lasted questionned in October 1982. Obviously, any change might be due to factors other than the television programme, and for this reason we split the sample into three groups, depending on whether they had seen all or part of the programme, had heard about it, or had neither seen nor heard about it. Of the 791 with complete data on these measures, 131 (17 per cent) had viewed at least part of the programme, a figure broadly consistent with national audience research statistics. Comparing the mean scores on the single-item measure and the eight-item index before and after the programme, we found small but significant changes in a more antinuclear direction on both measures for those who had viewed the programme, a marginal change for those who had heard about it, and no change for those who had neither seen nor heard about it. However, from the means of the before measures, there is some indication that those with more (negative) concerns about nuclear power were more likely to have been aware of the programme and to actually watch part of it. Even so, further analyses showed that, among those who had either seen or heard the programme, any attitude shifts were generally in an antinuclear direction, even for people whose prior attitudes were relatively favourable.

The persuasive impact of the programme is even clearer from measures of specific beliefs. Table 6.1 shows the means for the three viewing groups on beliefs about the possible effects of Sellafield and the causes of cancer incidence in the area. Those who had seen any of the programme were most (and those unaware of the programme, least) likely to think that the Sellafield plant had caused cancers locally and further afield, that the incidence of cancers around Sellafield was

148

excessive, and that any such excess was most probably due to routine waste discharge and to accidents at the plant.

We next looked at the ratings of danger from plutonium discharge and radioactive pollution. These allowed for comparisons between perceptions of the Sellafield reprocessing plant and the (much nearer) power station at Hinkley Point. Included in our sample were 141 residents of villages in the immediate vicinity of Hinkley Point (included in 'group X' as described in Chapter 2).

Table 6.1 'Nuclear Laundry' Study: beliefs about the reprocessing plant as a function of having seen the programme

	Seen the programme	Heard about it	Neither	F_{linear}
Beliefs[a]				
Plant caused cases of leukaemia and other cancers	2.1	2.4	2.7	47.83**
Incidence of cancers higher around Sellafield than elsewhere	2.2	2.5	2.8	31.10**
Plant caused cases of leukaemia in Scotland and the Hebrides	2.6	2.9	3.0	20.93**
Causes[b]				
Natural background radiation	3.2	3.3	2.9	3.38
Routine discharge of radioactive waste	6.0	5.1	4.6	29.11**
Accidents at the plant	6.1	5.1	4.8	19.23**
Other unknown factors	4.7	4.2	4.0	8.15*
n	131	277	383	

[a] Scores could range from 1 ('yes definitely') to 5 ('definitely not').

[b] Scores could range from 1 ('no effect at all') to 9 ('extremely strong effect').

**$p < 0.001$ *$p < 0.005$

Since we were particularly interested in whether people would draw generalized inferences about their *personal* levels of risk, we distinguished these *nearer* residents from those living *further* from Hinkley Point, so that, with the three viewing groups, we had a 2 x 3 factorial design. After controlling for the effects of having viewed or not viewed the programme, the nearer group was significantly less antinuclear than the *further* group, as measured by the attitude index (means, 31.8 versus 29.6), the single-item measure (4.4 versus 3.6) and the measure of support for a new nuclear power station in the South-West (3.7 versus 3.1).

Ratings of the extent of plutonium discharge or other pollution were in terms of four categories ranging from 'none at all' (scored as 1) to 'enough to represent a serious danger' (4). A separate 'don't know' category was provided. Subjects responding 'don't know' were excluded (there were more of these among those who had neither seen nor heard about the programme). As regards ratings of Sellafield, the *nearer* and *further* groups did not differ in terms of ratings of the extent of plutonium discharge (2.9 versus 3.0) but there was a fairly small, though significant, difference in ratings of pollution from Sellafield generally (2.7 versus 2.9). When asked the same questions about Hinkley Point, the *nearer* group gave considerably less pessimistic ratings than the *further* group: for ratings of plutonium discharges, the means were 2.2 versus 2.7, for radioactive pollution, 2.2 versus 2.6.

Thus, the dangers associated with Hinkley Point were generally considered to be less serious than those associated with Sellafield, and especially so among the nearer group. Even so, these data suggest that local residents based their inferences about the safety of Hinkley Point to some extent on generalizations from the evidence about Sellafield, as well as on broader attitudes towards any aspect of nuclear power. This is implied most clearly in answers to the question about plutonium discharges into the sea. Such discharges are technically impossible in the case of Hinkley Point. From the perspective of the industry, fears of *any* such discharges could appear 'irrational'. However, to ordinary people for whom the niceties of distinctions between one kind of nuclear plant and another would normally seem to have little day-to-day significance, evidence that nuclear pollution can occur in one context will appear a reasonable basis for assuming that it can also occur in another.

Conclusions from the 'Nuclear Laundry' study
Our data demonstrate measurable differences in general attitudes about nuclear power, and in specific beliefs about the Sellafield plant, as a function of whether or not people saw or heard about the television documentary. This fits in with more general impressions that the allegations made in the programme constituted a damaging blow to the public image of the BNFL operation at Sellafield. With media attention focused on the plant, further accidents attracted yet more publicity. Additional cases of leukaemia in Seascale and other communities nearby continued to receive coverage in later television reports. Other 'human interest' news included reports of houses being sold for around half their market value because of buyers' fears of pollution from the Sellafield plant. The causal connection between the plant and the excess cancer incidence remained a matter of scientific debate, but the very fact that the debate continued meant that the issue was kept alive.

The damaging effects of such publicity were not restricted to BNFL. Just as with the accident in the United States, the problems with Sellafield became seen to some extent as symbolic of problems with the British nuclear industry in general. Such symbolism acquired its own imagery or, in the technical language of psychology, its own cognitive accessibility. Pollution from Sellafield became, in other words, one of the first and easiest things for people to think about when asked any question about nuclear power.

The impact of Chernobyl

Serious though they were, the accident at Three Mile Island and the series of mistakes and mishaps at Sellafield do not begin to compare with the disaster at Chernobyl in terms of the devastation of the local environment and the risk to health from global dispersion of the radioactive fallout from the wrecked reactor. As we saw in Chapter 1, an antinuclear shift in attitude was evident in many countries in the months that followed. In our own research, we were able to look in sharper focus at how specific beliefs and evaluations were modified by news of the accident.

The Chernobyl reactor exploded on 26 April 1986. As mentioned in Chapter 4, at that very time we were collecting data concerning attitudes to oil and nuclear developments from residents of the town of Wareham in Dorset, which lies equidistant between the UKAEA nuclear research station at Winfrith and the BP oil field at Wytch Farm. As in our other studies, we were following the procedure of mailing questionnaires to names selected at random from the electoral register and asking individuals to give their names and addresses when they replied so that they would not be bothered by a reminder. Using this procedure, we had received 356 responses (from 913 deliveries) by 21 April, when non-respondents were sent a reminder letter and a second copy of the questionnaire. A further 124 replies were received in this second wave.

One section of the questionnaire asked for ratings of approval of actual, proposed or hypothetical local developments, following the wording of our previous studies. Inspection of the means for the two response waves suggested that later respondents were somewhat more antinuclear, especially when asked about their attitudes to building nuclear power stations elsewhere in the country, but showed no differences in their attitudes to oil or other industrial developments. This might have reflected an impact of the news from Chernobyl, but it is impossible to be certain. Although we can say for sure that the first wave responded before Chernobyl, we cannot be sure how many of the second wave had heard the news of the accident before they completed their questionnaires. Any questionnaires delivered quickly and completed immediately after delivery would still have been too early to be affected by the news. Moreover, even if all our later respondents had heard the news, they may have differed from earlier respondents in terms of some relevant factor. Perhaps the later respondents were more antinuclear even before the accident. If so, perhaps the accident prompted them to respond to our questionnaire but did not actually change the opinions that they would anyway have expressed. Even so, it is conceivable that the attitudes measured in our questionnaire would have been influenced, if news of the accident had been known. For this reason we planned a quick follow-up study, including questions dealing specifically with the Chernobyl accident (see Eiser, Spears and Webley, 1989).

152

Attitudes before and after Chernobyl

The sample for this study consisted of the 356 individuals who had responded by 21 April in the first wave of our 1986 study in Wareham, Dorset. All these were sent a further questionnaire in early June: 135 responded. The response rate (38 per cent) is lower than in our other studies, but this may be accounted for partly by the fact that no reminder was sent (it is practically the same percentage as responded without a reminder to the original mailing in April). The short time since the previous questionnaire was completed may also have been a disincentive. The 135 respondents included 96 men, 38 women and one person who did not declare his or her sex. This sex-ratio was not significantly different from that among non-respondents to the follow-up, who included 134 men, 81 women and 6 who did not declare their sex. Respondents, however, were significantly older on average than non-respondents (means: 52.7 vs. 45.3 years, $p <$ 0.001). Differences between the two groups in terms of their attitudes to nuclear power before Chernobyl were negligible.

The questionnaire contained various sections, the first of which consisted of sixteen attitude statements, with which individuals had to rate their agreement on a scale from 'very strongly disagree' (1) to 'very strongly agree' (7). These were treated as a Likert scale, with scoring reversed on eight of the sixteen items which expressed antinuclear opinions (2, 6, 7, 8, 11, 12, 13 and 14). The text of the statements is given in Table 6.2.

As will be noticed, eight of these (2, 3, 5, 8, 9, 12, 13 and 16) were derived from the eight-item scale included in the various stages of our 'four counties' study (see Chapter 2) as well as the 'Nuclear Laundry' study described earlier in this chapter. However, item 3 was modified from its earlier, more general but less reliable form ('overall, science and technology') and item 8 now referred to 'another major accident' rather than 'a major accident'. The eight new items were chosen to deal specifically with Chernobyl, and/or with attitudes towards the British as distinct from the Soviet nuclear industry.

The same sixteen statements were repeated in the next section, but responses of a different form were required. Instead of rating their *own* level of agreement, respondents were asked to 'describe the kind of opinion expressed by each statement (or the kind of person who could have made it)' in terms of five 9-point judgment scales. These were labelled: (a) 'not at all — extremely pragmatic'; (b) 'not at all — extremely alarmist'; (c) 'not at all — extremely concerned'; (d) 'not at all — extremely complacent'; and (e) 'extremely antinuclear — extremely pronuclear'.

Table 6.2 Attitude statements concerning Chernobyl

1. Compared with the dangers of other kinds of pollution, people's fears of the effects of radiation are out of all proportion.
2. Alternative technologies such as solar, wind, or wave power are a far more sensible investment than nuclear power.
3. Overall, nuclear technology has greatly improved ordinary people's quality of life.
4. The British nuclear industry can be justifiably proud of its safety record.
5. Nuclear power stations are far cleaner than any other kind of power station.
6. I feel sure that there are going to be many more nuclear disasters before very long.
7. What happened at Chernobyl could easily happen at any nuclear power station.
8. Even scientists have little idea what would happen if there was another major accident at a nuclear power station.
9. Nuclear energy is vital to Britain's economic future.
10. It is extremely unlikely that there will ever be another accident as serious as that at Chernobyl.
11. There's nowhere in Europe that's safe from the effects of a nuclear accident.
12. Britain should abandon all plans to build any more nuclear power stations.
13. Nuclear energy is far less economical than its supporters claim.
14. We've no reason to suppose that our nuclear power stations are any safer than those in the Soviet Union.
15. Nuclear power stations in Britain are built and operated with safety as the top priority.
16. Nuclear energy is the only practical source of energy for the future.

The next question asked 'Other things being equal, how *dangerous* or *safe* do you think it is to live at the following distances away from a nuclear power station?' This required ratings, on a scale from 'extremely safe' (1) to 'extremely dangerous' (9), of each of ten distances increasing geometrically from 1 km (or 5/8 mile) to 512 km (or 320 miles). Thereafter, respondents rated how much *attention* they had paid to the news about Chernobyl, and how *frightened* they had been by the news, on a scale from 'not at all' (1) to 'very much' (9). The final section of the questionnaire was repeated from that distributed in April to measure possible before-after shifts in attitude. This asked for ratings of approval, in terms of seven categories from 'very strongly opposed' (1) to 'very strongly in favour' (7), of the following actual, proposed, or hypothetical developments:

   (i)   The existing nuclear power station at Winfrith.
   (ii)   A new nuclear power station at Winfrith.
   (iii)   Building more nuclear power stations elsewhere in the UK.
   (iv)   The existing oil wells around Wytch Farm.
   (v)   New oil wells around Wytch Farm.
   (vi)   Inshore oil drilling elsewhere in the UK.
   (vii)   A new coal-fired power station in your neighbourhood.
  (viii)   A new chemicals factory in your neighbourhood.
   (ix)   Any other new large factory in your neighbourhood.

We first looked for any changes over time in approval of these nine industrial developments. Shifts in a more unfavourable direction were evident for the three nuclear items (mean before-after changes were respectively 0.33, 0.16, 0.34). Attitudes to coal (then the main alternative to nuclear power) appeared more favourable on the later measure (mean change −0.25), whereas the remaining items showed little change. Averaging over the three nuclear items, the mean change was 0.28, whereas that for the six non-nuclear items combined was −0.08, a highly small but significant ($p < 0.001$) difference. These data also showed that women were reliably ($p < 0.001$) more opposed than men to industrial developments generally, both before and after Chernobyl.

Agreement with attitude statements

We next considered the ratings of agreement with the set of sixteen attitude statements included for the first time in the *after* questionnaire. The Likert measure derived from these items had a very high internal consistency (0.96). In other words, the different items showed high intercorrelations, so that attitudes to nuclear power in general and to Chernobyl in particular went closely together. Thus, if someone expressed more negative evaluations of nuclear power in general, they also interpreted the specific implications of Chernobyl in more negative terms. Put slightly differently, those individuals who were generally more pronuclear tended to view Chernobyl as less catastrophic.

This overall attitude measure correlated significantly ($p < 0.001$) with approval of the three nuclear developments (i to iii) combined, both before and after Chernobyl. It also predicted before-after shifts in approval. Those individuals who were more pronuclear on this measure exhibited less shift towards disapproval of the three nuclear developments ($p < 0.001$). These more pronuclear individuals also shifted more towards disapproval of the six non-nuclear developments combined ($p < 0.01$) and of a new coal-fired power station in particular ($p < 0.001$). Ratings of how dangerous it was to live at various distances from a nuclear power station were averaged to give a single measure. This was closely associated with more negative attitudes on the same sixteen-item Likert scale ($p < 0.001$). More negative attitudes also went together with having paid more attention to the news about Chernobyl and being more frightened by the news.

Attitudinal judgments

The section of the questionnaire requiring descriptive ratings of the attitudes expressed by the sixteen statements relates to a broader theoretical area which requires a brief explanation. The field of 'social judgment' or 'attitudinal judgment' is concerned largely with how individuals assess *other* people's opinions, and with how such assessments are influenced by individuals' own attitudes on an issue. Early work in this field (Hovland and Sherif, 1952; Sherif and

Hovland, 1961) was built around the notion that individuals may 'assimilate' or 'contrast' other people's attitudes to their own, depending on whether or not the statements expressing such attitudes fell within a range of viewpoints which they still found acceptable.

More recently, attention has been directed to the role of the response language in terms of which such assessments are made. An approach termed *accentuation theory* (Eiser, 1984, 1990; Eiser and Van der Pligt, 1982) deals with the judgmental effects of the 'value connotations' of the response language. Simply stated, some adjectives that could be used to describe an attitude would normally be taken to imply approval of that attitude, that is to say, the use of such adjectives in normal conversation would normally be associated with a more positive evaluation of the other person or his or her attitude. To describe someone as holding a 'concerned' attitude thus implies approval rather than disapproval. Terms such as 'concerned' would therefore be said to carry 'positive value connotations'. Similarly, describing someone's attitude as 'alarmist' would be a way of conveying disapproval and such terms could be said to carry 'negative value connotations'.

The main prediction of accentuation theory is that, when judging others' opinions, individuals will prefer to use adjectives that are consistent, in terms of such value connotations, with their own approval or disapproval of the opinions expressed. Someone whose own opinion is pronuclear will prefer to describe an antinuclear opinion statement as 'alarmist' rather than 'concerned', because the more negative connotations of the term 'alarmist' convey his or her personal disapproval more adequately; this same individual should likewise prefer to describe a pronuclear statement as being 'pragmatic' rather than 'complacent', since the former term is presumably more positive than the latter. Contrary predictions can be made for individuals with antinuclear opinions, since they would not wish to describe opinions similar to their own as 'alarmist' but would be happy to describe pronuclear statements as 'complacent'. Such linguistic preference, according to accentuation theory, is reflected in more extreme or 'polarized' ratings on any given rating scale; more extreme ratings are given when individuals can use a positive term to describe their 'own side' on an issue, and a negative term to describe their opponents.

157

The prediction of accentuation theory tested here, then, is that more pronuclear individuals would give more polarized ratings than antinuclear individuals would on the 'pragmatic' and 'alarmist' scales, but that the reverse should be true on the 'concerned' and 'complacent' scales. In line with previous research, polarization was operationalized by calculating the mean difference between statements in the pro and anti halves of the scale. A greater mean difference between the two item groups (here, the eight pronuclear and the eight antinuclear statements) occurs when the statements are rated towards the opposite extremes of a given response scale. So, difference scores (based on division of the sixteen items into two groups) were calculated for each of the judgment scales. The direction of scoring was standardized so that, on each scale, evaluatively positive descriptions of pronuclear items and/or evaluatively negative descriptions of antinuclear items would lead to more positive (i.e. above zero) difference scores. Thus higher 'polarization' scores should correlated positively, on each scale, with more pronuclear scores on the overall attitude measure. This prediction was supported. Summing the polarization scores for the four scales together yielded a measure which, as predicted, correlate significantly with overall attitude ($p < 0.001$). There was no relationship between overall attitude and polarization of ratings on the fifth (antinuclear – pronuclear) rating scale.

Conclusions from the 'Before-After Chernobyl' Study.
The historical coincidence of the Chernobyl disaster with our research programme allowed a rare opportunity to measure changes within the same individuals in attitudes towards nuclear and other industries. Although the sample was comparatively small, it consisted of residents of a community with experience of a (small) nuclear plant a short distance from their homes. The issue of nuclear power was therefore not an abstract one for them. The news of Chernobyl led to the endorsement of more negative attitudes towards the existing nearby plant, the prospect of a new plant locally, and towards the policy of further nuclear development elsewhere in the country. With the possible exception of a shift towards more approval of coal, attitudes towards non-nuclear developments were unaffected.

Although it is unsurprising that an antinuclear shift occurred, the nature of the shift observed becomes more interesting on closer inspection. Firstly, the changes were in attitudes towards the nuclear industry locally and elsewhere *in the UK*. In other words, *any* shift in attitude involves a form of generalization, or of integration of the Chernobyl news into a broader category of knowledge about nuclear power, which could then be brought to bear on evaluations of *local* nuclear issues. The high correlations observed between these specific evaluations and the other attitude measures included in the *after* questionnaire further suggests that most individuals were evaluating Chernobyl and the local plant at Winfrith as though they were different examples of the same broad category of 'things nuclear'.

Secondly, although the shifts were statistically reliable, they were far from huge in absolute terms. We are looking at adjustments of previous positions, not large-scale conversions. One almost wonders how much more serious the Chernobyl accident would have had to be to produce large shifts on our measures. To cast this in perspective, approval for a new power station at Winfrith dropped from 2.9 to 2.8 on a 7-point scale (with men showing no average shift at all). In our 'Nuclear Laundry' study, approval (on an equivalent scale) for a new power station at Hinkley Point among those living nearer to the existing plant was 4.0, on average, among those who were unaware of the publicity about Sellafield and 3.2 among those who had seen all or part of the programme themselves. It is almost tempting to suggest that the Sellafield news had the greater impact, although any direct comparison of such figures would be highly speculative.

This underlines something very important about attitudes: they are very resistant to fundamental change. In both the studies reported so far in this chapter we found signs of shifts in attitude attributable to different kinds of 'bad news'. It seems that such 'bad news' added into a broader knowledge-base about nuclear power and, through a process of generalization, led to more negative evaluation of nuclear power in other specific contexts. However, this same use of a broader knowledge-base can dampen the impact of any single event. This is because this knowledge-base, and the evaluations to which it give rise, may influence how that event is *interpreted* (as well as affecting people's willingness to seek information about that event, perhaps). Even an event as striking as Chernobyl is open to different interpretations and can provide the basis for drawing different kinds of conclusions. Opponents of nuclear power were even more certain afterwards about what they already knew — nuclear power was 'bad'. Supporters of nuclear power, on the other hand, may have interpreted the Chernobyl accident in terms of other things they 'knew' — that, in many situations, nuclear power can be 'good'. In other words, the more favourable their broader attitudes to nuclear power, the more favourable was their interpretation of what Chernobyl implied.

Resolving inconsistency

Another way of saying that attitudes resist fundamental change is to propose that people adopt strategies to *resolve inconsistency*, a theme common to all cognitive consistency theories of attitude organization and change. Abelson (1959) outlined four ways in which people may resolve what he called 'belief dilemmas', in particular, states of uncertainty resulting from new information challenging their existing beliefs. The four strategies or 'modes of resolution' were called *denial, bolstering, differentiation* and *transcendence*. What is important about all of these is that they allow people's existing beliefs to remain more or less intact. In the context of Chernobyl, *denial* could take the form of a refusal to acknowledge the scale of the disaster, *bolstering* could take the form of stressing the compensatory benefits of nuclear power, and *differentiation* could take the form of insisting that 'our' power stations were built and operated in a 'totally different' way. *Transcendence* is defined by Abelson as the restructuring of cognitive elements at a superordinate level. This could include any claim that Chernobyl was some kind of 'blessing in disguise' for the nuclear cause (e.g. by being a mistake from which one could learn how to make nuclear power 'even safer', or by being the kind of 'worst case scenario' which many had feared could have been even worse, but which would now be unlikely to recur).

Such themes are implicit in a number of the attitude statements which differentiated supporters and opponents of nuclear power in our study just described. They also closely resemble arguments put forward by representatives of the British nuclear industry in the aftermath of Chernobyl. The following are examples suggestive, respectively, of the four strategies described above:

'The average individual dose in the UK resulting from Chernobyl, integrated to 50 years, represents about a week or two at normal background dose rates or the equivalent to having a three week holiday in Cornwall' (Gittus, 1987, p. 9).

'Nevertheless, in return for cheaper electricity and a reduction in the pollutants from conventional large scale power generation, the risk from the normal operation of nuclear power is exceedingly small' (Corbett, 1987, p. 21).

'It is clear that the Chernobyl reactor had features that would not be acceptable in this country and that there was gross malpractice by the station operators' (Saunders, 1987, p. 21).

'But if analysis of all aspects of the Chernobyl disaster leads to a better understanding of the causes of nuclear accidents and so to improved safety in the nuclear industry worldwide, we in the UK may well have reaped a substantial net benefit in the long run' (Corbett, 1987, p. 23).

Coping with decision conflicts

Similar issues arise in another line of research which considers individual differences in the strategies people use to cope with decision conflicts and uncertainty. According to Janis and Mann (1977), making decisions and resolving uncertainty can frequently be personally stressful. Some people, faced with new information that challenges their existing beliefs will show *unconflicted adherence* to their prior viewpoints as though they are simply not bothered by such a challenge. Others, however, will adopt a number of more or less effective coping strategies to deal with such conflict. Of particular importance is the strategy of *defensive avoidance*, which is expressed in various strategies of refusing to think about the problem or letting other people make the relevant decisions *(buckpassing)*, putting off any commitment *(procrastination)* and distortion or denial of the issues *(rationalization)*. Other patterns identified include *hypervigilance* (panicky or impulsive action on the basis of partial information) and *vigilance* — arguably the most adaptive — which is reflected in a weighing-up of benefits and costs of different options. More recent research has included consideration of individuals' evaluations of their own competence as decision-makers, an assessment that seems to relate closely to reliance on a strategy of vigilance (Burnett *et al.*, 1988; Radford *et al.*, 1986).

Work on Janis and Mann's (1977) *conflict theory* stemmed originally from a concern with the impact of fear-arousing communications. It seemed to us, therefore, that it might throw light on different patters of reactions to the news from Chernobyl, which was clearly fear-arousing for very many people. We therefore planned a final study, with the help of colleagues in different countries (Eiser, Hannover, Mann, Morin, Van der Pligt and Webley, 1990). This considered the reactions to Chernobyl among young adults as a function of the following factors: individual differences in preferred strategies of coping with decision conflicts in a political context; attitudes towards nuclear issues and policies in their own countries; and more general attitudes relating to politico-economic ideologies and the functioning of their national economies. Interest in this last factor relates to findings that support for nuclear power tends to be associated with endorsement of more right-wing or economically conservative outlooks, typically reflected in greater optimism about (long-term) economic performance (Webley *et al.*, 1986; Webley and Spears, 1988). The fact that the study was conducted in different countries (Australia, France, Netherlands, Germany and England) also permitted cross-national comparisons, possibly reflecting differences in the subjective immediacy, as well as geographic proximity, of the accident as well as in the shape of any national debate on nuclear issues. Such comparisons, however, are essentially uncontrolled.

Cross-national study

The sample for this study consisted of 840 young adults (385 men and 455 women) with a mean age of 23.6 years, recruited from the following cities: Adelaide, Australia ($n = 291$; 109 men, 182 women; mean age = 22.9 years); Aix-en-Provence, France ($n = 151$; 50 men, 101 women; mean age = 23.8 years); Amsterdam, the Netherlands (n

162

= 100; 64 men, 36 women; mean age = 21.9 years); Berlin (West), Germany (n = 143; 82 men, 61 women; mean age =29.0 years); and Exeter, England (n = 155; 80 men, 75 women; mean age = 20.5 years). Participants were mainly university students in psychology or social sciences, recruited on a voluntary basis during lectures or other teaching periods. There were uncontrolled differences between the groups in the timing of the data collection, which might have led to effects due to differences in the historical immediacy of the accident. The dates of testing were: Adelaide, July 1986; Amsterdam, February 1987; Aix, March 1987; Berlin, January 1987; Exeter, October 1986.

The questionnaire contained a number of sections which were common to all groups (translated as appropriate).

Political decision-making scale. This contained 12 items adapted from the Flinders Decision Making Questionnaire (Mann, 1982), a measure of individual differences in preferred style of coping with decision conflicts. These were reworded so as to focus on *political* decision-making. The text of these items is given in Table 6.3. Each item was rated 'true for me' (2), 'sometimes true for me' (1) and 'not true for me' (0).

Economic expectations. Participants rated how optimistic they generally felt about their country's economic performance, firstly over the next year, and then over the next 10 years, on a scale from 'very pessimistic' (1) to 'very optimistic' (9).

Reactions to Chernobyl. These items were taken directly from our previous 'before-after Chernobyl' study. Included were the questions concerning safety at different distances from a nuclear power station, about attention to the news from Chernobyl and about being frightened by the news. Participants also indicated their agreement with four statements selected from the sixteen used in the previous study (statements 1,6,7 and 10 in Table 6.2). Responses were on the same 7-point scale as previously used, from 'very strongly disagree' to 'very strongly agree', scored so that higher scores represented agreement with a more pronuclear position.

Self-ratings. Participants rated themselves on three 9-point scales (each scored 1 to 9), labelled: 'extremely antinuclear' – 'extremely pronuclear'; 'extremely ill-informed about nuclear issues' – 'extremely well-informed about nuclear issues'; 'not at all interested in nuclear issues' – 'extremely interested in nuclear issues'.

Other questions were phrased in ways relevant to the different national contexts. These included more or less direct indications of political preference, and items relating to different aspects of nuclear policy relevant to the country in question (to which participants had to rate their attitude on a 7-point scale from 'very strongly opposed' to 'very strongly in favour'). These typically dealt with the construction of nuclear power stations in the same region and/or country. The Adelaide sample, without experience of nuclear power generation in Australia, were asked instead about the topical local issue of the resumption of uranium mining in South Australia. The European samples were also asked about national aspects of nuclear waste disposal and reprocessing. Military aspects of nuclear power were also mentioned, including cruise missiles, and in the case of the French and British samples, independent nuclear deterrents. The Berlin sample was also asked about civil and nuclear power in what was then East Germany.

The twelve items in the political decision-making scale were submitted to a principal components analysis on the total sample. This allowed us to classify the items into four factors, as summarized in Table 6.3. The first of these reflects what Janis and Mann (1977) call *defensive avoidance*, the second combines *self-esteem* and *vigilance*, the third represents *passivity*, whereas the fourth seems closest to *hypervigilance*. Scores on each of these factors were computed for each individual, and the means for men and women and for the five different samples were compared. Women were significantly lower than men in terms of self-esteem/vigilance but higher in terms of passivity. There were a number of significant differences between national groups. In particular, Berlin and Amsterdam were lowest, and Adelaide and Exeter highest, on defensive avoidance.

A principal components analysis performed on the remaining sections of the questionnaire common to all groups allowed these data to be reduced to three (factor) scores per individual. A *pronuclear* score reflected principally self-ratings of attitude, the four-item Likert scale (measuring agreement with arguments that 'played down' the importance of Chernobyl), and saying that one had *not* been frightened by the news. An *involvement* score was based mainly on saying one had attended more to the news, was more frightened, and was more informed and interested in nuclear issues. The two items concerned with *economic optimism* yielded a third score.

Table 6.3   Items measuring decision-making style, classified according to loadings on each of four orthogonal factors

Factor I — Defensive Avoidance

I avoid thinking about political issues

I find politics boring

I prefer to leave political decisions to other people

There's nothing I can do that makes any difference politically

I put off taking a stand on political issues

I feel so discouraged that I give up trying to decide what stand I should take on political issues

Factor II — Self-esteem/Vigilance

I feel confident about my ability to make decisions on political issues

I think that I am a good judge of political issues

I like to consider all of the alternatives before making a political choice

Factor III — Passivity

I feel I should be more involved politically

It is easy for other people to convince me that their political opinion rather than mine is the correct one

Factor IV —Hypervigilance

I get very worked up when I have to make a political choice

The group means are shown in Table 6.4. As can be seen, men were more pronuclear than women and the Berlin sample was especially antinuclear. Amsterdam and Berlin were especially high, and Adelaide low, on involvement, and the economic optimism factor particular distinguished Amsterdam and Aix.

We then examined how these three attitude scores were related to the four decision-making style factors. Although being pronuclear was related somewhat to higher defensive avoidance and lower passivity, it was the involvement scores which were best predicted from measures of decision-making style ($p < 0.001$). Higher involvement was related to lower defensive avoidance, higher self-esteem/vigilance and lower passivity.

Table 6.4   Group means of attitude factor scores (x 100)

	Factor		
	I 'Pronuclear'	II 'Involvement'	III 'Economic optimism'
Males	13	16	12
Females	–25	–2	1
Adelaide	5	–21	–9
Aix	34	–6	–39
Amsterdam	3	34	88
Berlin	–64	33	6
Exeter	39	–5	–15

Note: Factor scores are standard normal deviates with a mean of zero and standard deviation of 1.

We next considered the statements dealing with nuclear issues of specific national relevance which differed between the various samples. These showed extremely high internal consistency in all five groups. One implication of this is that attitudes towards civil and military aspects of nuclear power did not produce differentiated patterns of response. Participants who opposed military use of nuclear power also tended to oppose its civil use. In all groups, there was a clear association between more right-wing political preferences and greater support for nuclear power.

Even so, within a context of predominantly antinuclear opinion, some types of issues gave rise to stronger opposition than others. Generally, military aspects were evaluated more extremely negatively than civil uses. Also there appeared to be a level of 'ingroup favouritism', in that existing nuclear plants or activities in one's own country were typically the most tolerated. For instance, the Adelaide group was most tolerant of existing uranium mines in South

166

Australia; the sample from Aix approved most of existing French nuclear power stations and the French nuclear deterrent; the Amsterdam group (while extremely opposed to underground waste disposal) was most tolerant of existing Dutch nuclear power stations; similarly, the Exeter group was most tolerant of existing nuclear power stations in the UK. The Berlin group was very opposed to nuclear power stations in the Federal Republic, but even more extremely opposed to any in East Germany (as it then was). In all groups, there was a clear association between more right-wing political preferences and greater support for nuclear power.

Conclusions from the cross-national study
This study does not permit any conclusions about the size of attitude shifts following Chernobyl (since no 'before' measures had been obtained), but it does indicate some of the variables which might mediate any changes that in fact occurred. The internal consistency of measures of attitude towards different aspects of nuclear power is striking, suggesting that specific beliefs — for or against — are largely shaped by broader ideologies or viewpoints accepted or rejected in their totality. One feature of this is that individuals whose own views were generally more pronuclear tended to 'play down' the importance of Chernobyl by denying that the same kind of accident could occur elsewhere, or that there would be many such accidents as serious as Chernobyl, and by asserting instead that the dangers of radiation were exaggerated. Such beliefs seem close to what Abelson (1959) might have described as differentiation and denial.

Support for nuclear power seemed quite closely related to a right-wing political orientation within the different national groups but was only rather weakly predictable from our measures of political decision-making style. The measures of style, however, were quite predictive of levels of involvement (e.g. attending to the news, and self-rated informedness and interest). Higher involvement went along with less reliance on the coping strategy of defensive avoidance and with greater self-esteem/vigilance. Thus, there would appear to have been individual differences, not simply in the implications people drew from the Chernobyl accident, but also in the extent to which people *tried* to draw any implications or to reappraise their existing beliefs.

167

One problem with the term defensive avoidance, which is borrowed directly from Janis and Mann (1977), is that it implies a deliberately motivated form of denial as a means of avoiding the negative arousal produced by the need to make a difficult decision. It is difficult to use such a term without implying a negative value judgment, although this is not our intention. As Earle and Cvetkovich (1990) point out, high scores on defensive avoidance often might be better regarded as reflecting lack of interest or involvement (as our more direct measures of these variables show), rather than 'defensiveness' in a strict motivational sense. The distinction is between not going out of one's way to gain information and deliberately going out of one's way to avoid learning (or thinking about) anything that could have implications for one's own behaviour.

Applying this distinction is difficult in practice, however, and depends on judgments such as whether the information is readily to hand and is likely to require personal changes in behaviour. Such judgments are not value-free. To say that ordinary citizens in countries of Western Europe (or even Australia) *ought* to think about the implications of Chernobyl would be to imply any of a number of controversial value judgments — that the events at Chernobyl have generalizable relevance to other countries, that Chernobyl actually confronts ordinary citizens with decisions which it is their responsibility to resolve, and that being politically involved is generally a 'good thing'. In short, it could be legitimate for someone to say of Chernobyl 'It's nothing really to do with me' or 'It's nothing I can do anything about', without being open to an accusation of motivated denial. This is rather different from, say, a heavy smoker claiming that information about lung cancer had no personal relevance, which is much more the kind of context in which Janis and Mann (1977) first developed the conceptual basis for their conflict theory (although even here the judgments are not value-free).

Even so, whatever one can infer about the resolution of motivational conflicts, our 'style' measures differentiated between individuals in terms of the ways they described their approach to political choices generally. Specific interest in Chernobyl thus seems to go along with an interest in politics generally. Once again, particular events are interpreted in terms of broader categories: if I'm not interested in politics and world events, I don't see Chernobyl as belonging to a category of problems requiring any deep thought or action on my part.

Conclusions

The feeling of danger and threat is a subjective judgment based on evidence. In this respect at least, there is no fundamental difference between the judgments made by lay people and by industrial experts. The important question is what counts as evidence, and how that evidence is interpreted and used. In the next chapter, one of the things we shall be looking at is how quantitative definitions of risk have permeated the industrial perspective and shaped national policy. Ordinary people's views of nuclear risk, however, do not typically involve any precise definition of how a nuclear accident could occur or deal with precise mathematical probabilities. Instead, the underlying reasoning seems to be steered by more general assessments of whether nuclear power generally is dangerous and/or badly managed. Evidence that nuclear plants can 'go wrong' in some respects is taken as a pragmatic basis for inferring that they could 'go wrong' in other respects too. Any kind of nuclear accident is taken to show that nuclear power cannot be totally safe, once one considers nuclear power as a totality. Furthermore, evidence that incidents occur with mild consequences is taken to show that they could also, with a little less luck, occur with severe consequences.

Should one consider nuclear power as a totality, or only look at the probabilities of specific accidents involving specific technologies? There is no single hard-and-fast answer to this question. From a strictly technical point of view, it is quite correct to say that evidence of the poor safety record of one piece of hardware has no necessary relevance to the risks associated with another piece of hardware. However, if risks are not completely definable in terms of hardware — if we need to add more general considerations (such as the potential for human error) — then the distinctions between one kind of nuclear plant and another may start to appear less crucial.

In any event, as far as ordinary people's opinions are concerned, it is clear that a good deal of generalization occurs. Incidents leading to unplanned but limited damage make the possibility of serious accidents seem greater. Serious accidents increase the subjective likelihood of even worse disasters. Ordinary people acquire a knowledge-base about nuclear power in terms of which catastrophes are far from unthinkable, simply because major accidents have occurred and been widely reported. It is therefore not all that surprising that bad news about nuclear power in one specific context will often be reacted to as bad news about nuclear power in general.

Yet it is important not to overstate the impact of nuclear accidents on public opinion. Generally, the changes constitute modifications rather than extreme conversions. Although typically highly reliable in direction, attitude shifts tend to be modest in absolute size. Furthermore, they tend to be more marked in the short than the long term. For instance, Renn (1990) presents opinion poll data which show marked increases in the numbers of opponents of nuclear power in the immediate aftermath of Chernobyl in each of eleven European countries, but a moderation of this effect by 1987. News of Three Mile Island, of Sellafield, or of Chernobyl, becomes integrated into people's knowledge-base. Such information is not ignored, but neither does it swamp everything else one knows about nuclear power. The more distant an accident becomes in time and (to some extent) in place, the less it is at the forefront of one's mind when one is asked to describe one's attitude towards nuclear energy or to assess the likelihood of a particular disaster.

Just as information about accidents becomes integrated with existing knowledge, so such knowledge will influence the way in which the information is interpreted. Those familiar with the intricacies of nuclear technology will have a more complex and articulated set of concepts in terms of which to try to understand what has happened, even if hard evidence is initially difficult to come by. Ordinary people will have to rely on what they are told, but if what they are told seems confusing, incomplete and contradictory, this is likely to increase not only fear but suspicion. New information will be interpreted differently, depending on its apparent reliability.

But not everything that ordinary people feel or know about nuclear energy is bad, by any means. Those living close to existing nuclear plants will have had day-to-day experience of nuclear safety and not just knowledge of nuclear dangers (although some will undeniably have 'inside' information about these too). Evidence that one's local plant is 'liveable-with' and news that a reactor in the Soviet Union has had its top blown off, taken together, demand some form of resolution of apparent inconsistency. It thus becomes entirely reasonable to look for factors that distinguish, say, the Chernobyl reactor from that at Winfrith, if Winfrith is something one knows a bit about. Different kinds of knowledge produce different kinds of attitudes towards nuclear power in general. Different knowledge and attitudes together are reflected in different categorizations, that is, judgments of similarity and difference among 'things nuclear'. These categorizations are both interpretative and evaluative, acknowledging the need for knowledge and attitudes to adapt, but often leaving people's fundamental beliefs and orientations intact.

Chapter 7

Danger, Trust and Public Policy

Our emphasis throughout this book has been on the structure of the attitudes of ordinary people, on the importance they attach to particular aspects of proposed developments, on their balancing of the benefits and costs, and on their estimation of the risks. These attitudes are not merely interesting in themselves. They set much of the context in which energy policy is formulated and implemented. They comprise a form of 'common sense' which, for better or for worse, is the basis of public acceptance and opposition. In one respect in particular, this 'common sense' can be strikingly at variance with the opinions of those working within the nuclear industry. Many ordinary people see nuclear power as very dangerous and have little trust in official statements of reassurance. It is not our place to say whether these people are right, nor even to estimate what percentage of the general public feel this way. It is enough that people with such fears are easy to find. Whatever the future of a nuclear energy policy, this leaves us with something to be explained. Why are official views on nuclear safety not universally accepted by ordinary people? Part of the answer lies in the impact of 'bad news' on the international image of the nuclear industry, as we saw in Chapter 6. Bad news may also be treated as more newsworthy by the media, as the findings in Chapter 5 suggest. However, another reason may be that the industry and the public may be employing rather different conceptualizations of risk from one another.

In this final chapter, we shall start by looking at the contrast between such 'expert' and 'lay' definitions of risk. We shall then return to the local context of our research project, describing how plans for a new nuclear power station at Hinkley Point were presented and appraised at a public inquiry. As will be described, many of the crucial points considered at the inquiry revolved around precisely the kind of conceptual and social psychological issues with which our research has been concerned. What is risk? What does it mean to say that a given level of risk is 'tolerable'? How does the willingness of local residents to tolerate a risk depend on their estimation of compensatory benefits? Is the remote possibility of catastrophe the most important disincentive to acceptance by local residents, or is the certainty of an irreversible change to the visual environment a factor which is regarded at least as seriously? What balance should be struck between local costs and national benefits? Finally, we shall conclude this chapter and this book by looking back over the main findings of our research, and looking forward to how social psychology may help inform the decisions about alternative energy futures that will need to be made in a world that remains as uncertain as ever.

Risk and uncertainty

Technological estimates of risk are expressed in terms of probabilities — probabilities of an accident happening at all and probabilities of the consequences of any accident reaching a specified level of severity. Levels of severity are frequently expressed in terms of numbers of fatalities among workers at a plant and/or members of the public. These calculations are derived from more complex estimates of the scale of possible radioactive escapes and emissions, and of their potential for causing disease, particularly cancer. Even so, the emphasis on fatalities, though appealing to those looking for a neat quantitative index of severity, tells only part of the story. To provide a fuller picture of the consequences of a nuclear catastrophe, one would have to take into account other impacts too, such as the incidence of non-fatal illness, birth defects, as well as social disruption caused by evacuation and personal stress generally. No doubt these other impacts would tend to rise along with the number of estimated fatalities, but the correlation might well be far from perfect, depending on a variety of factors such as the nature of the accident, proximity to centres of population, and prevailing climatic conditions.

174

Despite such (generally) acknowledged difficulties with the measurement of severity, the debate over nuclear risk has tended to focus on probabilities, for instance of a melt-down, or of an escape of radioactivity into the atmosphere following some kind of fracture or explosion. Not surprisingly, the nuclear industry insists such probabilities are extremely low. Again not surprisingly, opponents of nuclear power are far more pessimistic. It is not for us to suggest who is right and who is wrong, but rather to try and make sense of this difference of opinion. Making sense of different opinions can typically involve showing how each are sensible, though in different ways.

Probabilities come in many shapes and sizes and not all are equally straightforward to interpret. Some of the simplest probabilities are those inferred, by extrapolation, from the frequencies of previous events. For instance, it is relatively easy to estimate the likelihood that the total number of traffic accidents nationally during next January will be greater or less than a given target figure. This is because we have have access to records of accident frequencies during previous Januaries and we can assume — albeit allowing for changes in traffic density and variable weather conditions — that such patterns will remain relatively stable over time. Most importantly, traffic accidents are very frequent events when considered on a national scale, even though the chances of any one individual being involved in an accident in any one month are very small indeed.

Extrapolation of this kind is much less secure statistically when we are dealing with very rare events. Major nuclear accidents have happened, of course, but they are thankfully still very rare. Moreover, when considering a very small sample of events, it may always be possible to identify some special condition (such as a different reactor design) which may weaken the relevance of such events as a basis for predicting what might happen elsewhere. Very much more often than not, estimation of the likelihood of nuclear accidents involves judging the probability of something that has never yet happened, at any rate under the particular conditions in question. How then should such judgments be made? One approach is to formalize one's assumptions concerning what it would take for an accident to occur, and estimate the probabilities of each step in a

175

particular chain of events. Since we typically (in this kind of context) have to extrapolate from a limited data base, this is necessarily an uncertain and inductive procedure, but induction does not need to be undisciplined. Formalization of this inductive approach is the basis of an extremely influential technique of risk analysis, known as probabilistic risk assessment.

Probabilistic risk assessment

Probabilistic risk assessment (PRA) involves identifying sets of 'initiating events' which could lead directly or indirectly to the occurrence of a particular accident or incident. A model is then hypothesized to describe how these events might follow on from one another to culminate in an accident. This model can be represented graphically in terms of a tree-diagram referred to as an 'event tree' or 'fault tree'. The probability of each component event (branch in the tree) is assessed or estimated and the overall probability of the final accident is calculated from the products of the component probabilities (following the rule that Prob.(A and B) = Prob.(A) x Prob.(B)). This assessment technique has proved extremely valuable in many industrial settings, especially in identifying weak links in safety procedures and quantifying the relative improvement in safety if additional precautions are introduced. It has the special advantage of making designers state explicitly what kinds of event sequences they expect to lead to accidents.

A need for caution in interpreting the results of PRA arises from the following considerations:

a) As an absolute measure, the calculated probability of an event sequence can never be more statistically reliable (less prone to error) than the probability of any one component event or branch. Just as the component probabilities are multiplied by each other, so are their associated error variances or confidence limits. This point is rarely acknowledged, but is crucial to comparisons of outcomes of PRA exercises with absolute (e.g. statutory) safety standards. Often, in complex sequences, the probability of some events cannot be reliably inferred from existing data and simply has to be guessed. The rule of thumb in such cases is to try to err on the side of caution.

b) The model may be incomplete, that is, it may leave out crucial paths which could also culminate in the accident. It is a matter of judgment whether a model is sufficiently complete to provide a basis for accurate assessment or at least for useful comparisons.

c) Related to the above, some kinds of contributory factors are notoriously difficult to fit into a PRA approach, either because their associated probabilities are difficult to assess or because their point of impact within the model is difficult to define. Much of what falls in the broad category of 'human error' is problematic in either or both of these respects.

d) The assumed independence of different contributory factors or events is crucial to any calculation. However, this independence cannot be assumed *a priori*. Just as plant designers attempt to guard against the simultaneous failure of different parts of the system, there is a need for better modelling of possible interactions between multiple causes, including varieties of human error.

Notwithstanding these reservations, PRA has been used as the basis for quantitative estimates of the probabilities of serious nuclear accidents. Since such accidents have been regarded as requiring a whole sequence of separately unlikely mistakes or failures to occur, the combined probabilities, as calculated, have often come out as vanishingly small. For instance, the chance of a loss-of-coolant accident in any given reactor in any given year was once estimated at 1 in 10 million, or 10^{-7} (Rasmussen, 1975). More recent estimates based on such techniques are higher, but still very small.

'Subjective' risk

Because such 'expert' estimates of nuclear risks (as well as of many other industrial hazards) are so low in absolute terms, it is tempting for advocates of nuclear expansion to argue that high levels of public concern are exaggerated and even 'irrational'. Essential to this

177

argument is a distinction between so-called 'objective' risk estimates (as revealed, for instance, by PRA) and 'subjective' judgments of risk made by ordinary, inexpert, members of the public. In simple terms, the argument is that people who fear nuclear power do so because they misperceive the risks already recognized (but calculated as minimal) by the industry.

There are many objections that can be raised to the introduction of an 'objective-subjective' dichotomy. If by 'subjective' we mean reliance on some form of inference that goes beyond the facts already known, then all risk estimates are subjective to some extent. Even when making predictions about something where we have plenty of previous data (e.g. road traffic accidents), we still need to make assumptions about the underlying statistical distribution of our data, the likely effect of random factors, and such like. Or again, there may strong 'objective' evidence that radiation can increase the risk of cancer, but many assumptions need to be made about the dependence of any disease process on whatever exposure levels over whatever periods of time among different age-groups or distinct populations. There is a big difference between saying that people make judgments of risk on the basis of uncertain or inadequate information, and saying that they process such information irrationally, as the notion of 'misperception' implies.

Nonetheless, the idea that risks — or more generally, probabilities — may be 'misperceived' derives support from a large body of experimental research on the psychology of decision-making. It is quite clear that people find statistical information and relationships difficult to interpret. Tversky and Kahneman (1973) coined the term 'cognitive heuristics' to refer to some of the decision-making strategies which people use when faced with the task of considering the likelihood of uncertain events. Reliance on such strategies can lead to a number of apparent errors of reasoning, depending on the kinds of information presented for interpretation (Nisbett and Ross, 1980). We have come across such heuristics (e.g. availability, anchoring and adjustment) in previous chapters. If we define 'rationality' as merely the following of some statistical algorithm, it is clear that the way we think about probabilities in general (and not just risks of disasters) exhibits many demonstrable signs of 'irrationality'.

As was illustrated in Chapter 4 (in the studies on comparisons of the contributions of different energy sources), variations in the context or format of particular questions can lead to marked differences in response. All this has led to a view that people's subjective judgments of probability and uncertainty are highly vulnerable to a range of contextual cues and biases. From this position it is a short step to claiming that lay people's fears of nuclear accidents are not only exaggerated, but the product of faulty reasoning.

So can public fears of extremely 'improbable' events be thereby dismissed as 'irrational'? One reason to hesitate before drawing such a conclusion is that we then are still left with the problem of explaining how human thought processes could have evolved selectively so as to yield regular and predictable errors of this kind, if such errors actually are maladaptive for ordinary living. The point is that such 'irrationality' is so defined solely on the basis of formal or analytic criteria, without reference to the consequences of making decisions in one particular way rather than another. In a real world where actual probabilities are in practice or even in principle unknowable, following the formal rules of mathematical statistics may not be high on the agenda. Indeed, it may even be impossible. The informal approximation to 'rationality' embodied in the use of cognitive heuristics may in fact make a good deal of pragmatic common sense (Hogarth, 1981; Eiser and Van der Pligt, 1988). Conversely, the insistence on precise statistics in a context where many factors are incalculable may itself be far from rational.

It is also difficult to insist on a purely mathematical definition of rationality when serious dangers are deemed to be involved. Again, if we think how our cognitive capacities have evolved, it may be better to go hungry for a while than to eat food suspected of being poisonous and it may be better to take cover at the approach of a predator even if the predator has more appetizing prey at hand. Avoiding unknown fungi or a swim in waters where sharks are sometimes seen may mean that one misses out on some rare delight, but at least it means that one lives to see another day. To adopt the rule 'Better safe than sorry' may lead to wasted opportunities and it may invite accusations of a lack of enterprise, but it cannot be dismissed as 'irrational' in contexts where dangers appear either

179

large or indeterminable. None of this accounts for why nuclear power plants should be seen by many people as especially threatening compared with most other kinds of industrial development. But what it does imply is that people who feel threatened, for whatever reason, are unlikely to be more than partly reassured by a few statistics. We do not tend naturally to think of danger in a formal statistical manner, so an (exaggerated) perception of danger cannot be simply attributed to a misperception of statistics.

Trust and responsibility

Another objection to defining risk in terms of probability is that it evades the issue of causality. Even if death from radiation from a nearby nuclear plant is as infrequent as death from lightning, there will be differences in how responsibility is attributed. It would scarcely be comforting to be told that plant operators and managers could not guarantee one's safety from radiation any more than one's safety from lightning! The distinction is that nuclear risks, however small (and however much smaller than many people may imagine), are a product of industrial activity and hence of human decisions. Lightning, on the other hand, is a chance event against which one may seek protection but which cannot be controlled. Public perceptions of nuclear risks are therefore bound up with perceptions of the activities and decisions which give rise to such risks. As such, they depend on a form of reasoning which is at least as social as it is statistical. In short, appraisals of nuclear power as safe or dangerous may depend ultimately on judgments of the competence and trustworthiness of responsible members of the industry, as well as of the reliability of the technology which they control. Nuclear safety depends on the quality of human decision-making at many levels from design to manufacture to management and operation.

One psychological approach to the assessment of decision-making quality is signal detection theory, originally developed to decribed performance on perceptual tasks demanding the discrimination of some signal from background noise (Swets, 1973). We do not need to go into the mathematical details of the theory as it applies to specific kinds of perceptual stimuli. It is sufficient just to grasp two features of its basic logic. The first is that one can characterize decisions in terms of the sensitivity of discriminations, i.e. the extent to which genuine 'signals' will be more likely to elicit a response than misleading 'noise'. The second feature is the setting of a response criterion so as to influence the proportion of different sorts of errors (specifically, 'false-negatives' or 'misses' as distinct from 'false-positives' or 'false alarms'). These two features interrelate, in that, if sensitivity (reliability of discrimination) is imperfect, an excessive number of misses can only be avoided at the price of an increasing number of false alarms.

To put this in the context of nuclear safety, the fundamental question is whether plant managers, operators (and even automatic detection instruments) can recognize sources of potential danger and respond appropriately. An appropriate response may be anything from a minor adjustment to a total shut-down (or even a decision not to build a particular kind of plant in the first place). A failure to respond to an actual source of danger (a 'miss') is obviously extremely dangerous and everything needs to be done to avoid anything like that happening. However, if real danger signals cannot be discriminated with 100 per cent accuracy, 'misses' can only be avoided at the price of a certain number of 'false alarms', or, in other words, by being more cautious than is strictly necessary in every circumstance. In fact, extremely sensitive detection procedures and a good deal of caution are built into the design and operation of modern nuclear plants. But suppose that, despite all of this, some members of the public do not believe that potential sources of danger have all been clearly anticipated or can all be clearly recognized. There would then be seen to be a high probability of 'misses' which could only be reduced at the price of a very high number of 'false alarms'; in fact, if discrimination sensitivity is seen as very low, the response criterion required to keep 'misses' to a minimum may need to be biased so far in the direction of allowing false-positive errors that it amounts practically to a total shutdown of the system (see Eiser, 1990, for more details on this argument).

Perception of nuclear risk is thus inseparable from the question of trust. If, for better or worse, the nuclear industry is seen as not being in complete control of its plant — whether because of a lack

181

of adequate expertise or attention or simply because nuclear energy is perceived to be too powerful to control completely — then nuclear power stations will be regarded as very dangerous, whatever the statistics suggest. Indeed, such a perception cannot be shown to be unreasonable on statistical grounds alone. Remember that the statistics in question are themselves no more than estimates, based on extrapolation from a limited body of directly relevant data and/or inferences from a selective modelling of possible routes to disaster (in the context of PRA). If the 'experts' have failed to take account of some special circumstance contributing to risk, or have underestimated the potential impact of human error or mismanagement, then the dangers will indeed be greater than the presented statistics imply. In other words, the quality of the 'expert' statistics will be evaluated in much the same way as the expertise of the industry itself. On the other hand, if such expertise is fully trusted, the same statistics will be seen as reassuring, or even just as confirming what is already known (that nuclear power is very safe). Another way of looking at this is that we often willingly put ourselves in the hands of 'experts', even though the statistical risk of a chance mishap or of even death are far higher. Such 'experts' can be anyone from a bus driver to a surgeon. Whatever the context, the question is one of trust. If we trust them to do their job properly, we tolerate the risk. If we feel they are incompetent (or drunk!), we do not get on the bus and we do not have the operation, and no quoting of comparative risk statistics will persuade us that we are being unreasonable.

Weighing the costs and benefits

Another very important reason why the question of tolerability of risk cannot simply be settled by statistics is that any risk, however small, is only submitted to wittingly in order to achieve some benefit. This is as true of nuclear power stations as it is of bus rides or surgical operations. How might the costs and benefits of nuclear power be evaluated against one another? As our own research has shown, much of the objection to the building of a new plant does not turn on whether the annual risk of an uncontrolled release of

182

radiation is 1 in 10,000 or 1 in 10 million. Rather, it relates often to less cataclysmic but far more certain consequences. The disruption of a quiet community and the compromising of coastal scenery may not exactly be life-threatening effects, but, if left uncompensated, they may be seen as constituting an unwelcome worsening of the quality of life of local residents. So the question of the likely benefits of a new power station are crucial for an understanding of public attitudes. As far as local attitudes are concerned, a remaining problem is that the main benefits will be felt by people living some distance from the immediate locality of the plant. Furthermore, the extra employment and business opportunities afforded by such a huge construction project will be enjoyed unevenly by different residents. New roads, shops and community centres maybe seen as mixed blessings by those who have opted for a less urbanized way of life. The situation could be very different if the construction site was in the industrial wasteland of some large city, but that is not where nuclear power stations, because of safety rather than scenic considerations, happen to be built.

Although direct financial compensation might, in principle, be offered to local residents, this is not a strategy that has historically been favoured by the British nuclear industry. Indeed, any 'hand-out' might appear to be an attempt at bribery which might only serve to increase distrust and local opposition. However, without extra compensation, the strictly local benefits of a nuclear plant are unlikely to be fully adequate (for existing residents) to balance the local costs. Much of the reason for this is that a construction project would not have to be nuclear in order to offer such benefits, but if it were not nuclear it would probably be seen as less dangerous and it would almost certainly be somewhere else. Ultimately, an undertaking of the scale of a new power station, let alone a programme of construction of a series of power stations, demands justification at least at the regional level and most probably at the national level. If that justification is seen as convincing, the dangers and inconveniences might be tolerated even by those residents who are personally most at risk. Without such justification, even small risks and inconveniences may be seen as intolerable. Tolerability and apparent equity are thus inextricably related.

183

The Hinkley inquiry: the legacy from Sizewell

The inquiry into the building of the planned PWR at Hinkley Point C finally opened in October 1988, with Michael Barnes QC, in the Chair as Inspector. A public inquiry is a peculiarly British institution with strictly circumscribed powers of decision-making but with the authority to make recommendations to the Secretary of State, having duly weighed the evidence on both sides and the expressions of opinion by local residents, experts and industrial representatives. The implicit ethos is not that of a search for scientific truth, but of protection of public interest and a fair balance of different interest groups, under the provision of various Acts of Parliament (such as, in the case of a local planning inquiry, the Town and Country Planning Act of 1971). Interpretation of the law thus features at least on a par with interpretation of scientific evidence and the inspectors who chair such inquiries, though typically assisted by technical experts (assessors) who may interrogate witnesses, are usually lawyers rather than scientists. Different parties to the inquiry may likewise be represented by legal counsel (if they can afford it), yet at the same time essentially anyone who feels they have something worth saying to the inquiry can come and say it, so long as they follow relatively simple protocols. This can make for an extreme variation in the kinds of 'evidence' the inspector is called upon to evaluate. Where at least one side has no interest whatsoever in allowing the inquiry to come to a quick conclusion, matters can carry on for a very long time indeed.

Imagine a cricket match, played for months on end, for the most part in front of only a handful of spectators, without an upper limit on the size of either team or even on the number of teams that can play at the same time, with amateurs mixing their skills with those of visiting professionals, and with no predetermined date by which the match should finish, and you will have some idea (even if you don't understand cricket!) of the confused mixture of tension and tedium which characterizes such events on a day-to-day basis. Yet the conclusions of an inquiry are not trivial. In that they purportedly carry the weight of legal, popular and technical opinion, somehow reconciled or balanced with each other, they can provide standards against which government policies and industrial practices can be held to account.

Despite the long delay since the choice of the Hinkley Point site, the CEGB had been far from idle. In the meantime, there had been another lengthy public inquiry into the building of a new power station at Sizewell on the Suffolk coast. This had been an especially important inquiry, since what was at stake was the building of a Pressurized Water Reactor (PWR), the first of its kind in Britain. The CEGB team entered the Hinkley inquiry in a spirit of optimism. The positive decision from the earlier inquiry at Sizewell made them confident that they had established the 'generic' case for the desirability and safety of a programme of new PWRs, so that all that would need to be talked about were issues related to the specific site under consideration. They did not see why they should have to go over old ground in establishing whether PWRs should be built at all. The questions for the inquiry, as they saw it, were simply whether one of the planned PWRs should or should not be built on this specific site and, if so, what special arrangements would need to be made to protect the local environment and avoid undue disruption to the local community.

There was one fundamental issue, however, that had been left unresolved by the Sizewell inquiry, which was the weight to be given to public opinion, particularly over issues of risk. Sir Frank Layfield, the inspector at the Sizewell inquiry, was concerned that information about nuclear risk and safety had not been presented in such a way that members of the general public could be expected to make an informed judgment about the risks associated with a new PWR. The industry might produce technical evidence sufficient to demonstrate to experts that these risks were very small indeed. However, such evidence by itself would not establish that such risks would be tolerated by members of the public.

The question of how much weight should be given to public opinion in planning decisions was muddled both before and after Layfield's report (O'Riordan, Kemp and Purdue, 1988). At one level, public opinion elects governments, and governments (if only by default) make strategic decisions about energy policy. In carrying out such decisions, governments depend considerably on the advice of inquiries which are so constituted as to allow different shades of opinion, particularly local opinion, to be expressed. But what happens, what should an inspector recommend, if expert and public opinion are on different sides? The government's own stand was clear. Margaret Thatcher in November 1985 dismissed opposition to new nuclear power stations and to the reprocessing plant at Sellafield as 'irrational and misguided'.

Layfield adopted a more conciliatory tone. If the public were uninformed about the extent of nuclear risks, they should be provided with information on which to base their judgments, in an intelligible and 'transparent' form. The issue of risk, however defined, should be central to any decision about whether or not to give consent for a new power station. Some levels of risk may be so high as to be 'intolerable' and, if so, consent should be refused. Part of what Layfield meant by tolerability is that 'the benefits are expected to outweigh the risk' (Layfield, 1987, Summary para 2.101c). Furthermore, whether or not a given level of risk is defined as 'tolerable' is in large part a matter of opinion. As Layfield (Summary, para 2.101h) put it, 'the opinions of the public should underlie the evaluation of risk'. Only too aware of how difficult it had been for him to take systematic account of public opinion, he recommended that the Health and Safety Executive (HSE) 'should publish a consultative document to enable public, expert and Parliamentary opinions to be expressed' (Summary, para 2.102f). The result of this recommendation was a document entitled *The Tolerability of Risk from Nuclear Power Stations* presented in evidence to the Hinkley inquiry on behalf of the HSE by John Rimington, the HSE's Director General.

The tolerability of risk

The HSE document was an attempt by a government regulatory body, not simply to calculate nuclear risks in quantitative terms, but to prescribe what quantified risk levels should be regarded as tolerable. However, in one important respect the HSE declined to take on the whole brief that Layfield had given them. As Rimington explained in his evidence (HSE1, 3.29):

186

'The Document on the Tolerability of Risk does not develop the general question of the economic benefits of nuclear power. This is not in my judgment a suitable subject for a regulatory body, which must of course have regard to the narrower question of the cost of its requirements in respect of particular installations or types of installation, but which cannot pronounce on the case for or against nuclear power other than on safety grounds.'

The HSE's insistence that a consideration of the benefits of nuclear power were beyond their proper terms of reference meant that the debate at the Hinkley inquiry took a strange form. Defensible although this position may have been, it obscured the simple point that any risk must be offset by benefits if it is to be regarded as tolerable. The CEGB were happy to leave this point obscure, since they had no wish to reopen the general question of whether building more power stations was a good thing. Thus the HSE examined the issue of whether nuclear risks were tolerable without estimating whether they were offset by any compensatory benefits. The question of cost and benefit was addressed only in a more restricted context, namely the assessment of whether risks at a nuclear plant had been reduced to the point that further improvements in safety would only be achievable by a disproportionate and impracticable increase in costs. This criterion is termed the 'As Low As Reasonably Practicable' (ALARP) principle.

So how could tolerability be evaluated without regard to benefits? The approach taken by the HSE was to offer a quantified assessment of risks from nuclear power stations and to compare these with risks associated with other industrial (and non-industrial) activities and installations. A distinction was drawn between assessment of individual risk on the one hand and of societal risk on the other. Individual risk is the proportionate risk of any single employee or member of the public dying from a specified cause, within any given year. For example, the individual annual risk of being killed in a road accident (averaged over the whole population of Great Britain) is calculated as around 100 per million, or 1 in 10,000. Societal risk refers to the estimated probability of some major disaster occurring in any given year, resulting in a large number of fatalities. For example, the pattern of previous air crashes in the Home Counties is used as the basis for a calculation of a probability of 1 in 100 million of an aeroplane crashing into a full football stadium in London in any one year. More tangibly, the annual chance of a major explosion at the oil refineries on Canvey Island in the Thames estuary, leading to 1500 deaths or serious injuries, is calculated to be as high as 1 in 5000.

So how do the risks from nuclear power stations compare against these figures? By applying the techniques of probabilistic risk assessment (PRA) as far as — and perhaps further than — available data permitted, the HSE document calculates the annual risk of death to individual employees exposed to relatively high doses of radiation to be 'in a range from rather higher than 1 in 10,000 to about 1 in 40,000' (p. 23). This corresponds roughly to the range from the risks in heavy manufacturing or mining to those in other manufacturing industry. Individual risk to members of the public living in the vicinity of a nuclear plant during normal operation, on average, 'will be well below 1 in 100,000 per annum' (p.24). The 'societal risk' of an accident at any given reactor serious enough to lead to the eventual deaths of 100 members of the public is similarly calculated at 1 in a million per annum. One needs to bear in mind that the dangers of such casualties from accidents or normal operation, considered over the country as a whole, depend on the number of power plants actually in operation, so that a network of twenty reactors will be associated with an overall risk that is twenty times more probable. Even so, the HSE's conclusion is that the risks posed by nuclear industry, both to employees and to members of the public, are well within the range of other 'tolerated' industries and activities, and well below a level that would be regarded as definitely 'intolerable'.

The argument underlying these conclusions can be challenged on a number of points of detail and even principle. The manner in which the possible effects of human error are considered is especially simplistic in terms of both behavioural and numerical assumptions (cf. Reason, 1990). Even more worrying though, is the attempt to define tolerability (to the public) in terms of technical assessments that by-pass any proper analysis of public opinion. It is, in short, a prescriptive definition. It rests ultimately on the view that, if the calculated risks associated with a new development are no worse than those of existing developments (or activities) which appear to be tolerated by the public, then it would somehow be 'unfair' to block the development on safety grounds. Alternatively, there may be an implication that the public not only should but will tolerate or accept the risks from a new nuclear power station. This is quite close to what Starr (1969) termed the 'revealed preference' approach, according to which the very existence of some installation or activity is assumed to 'reveal' that society has concluded that the associated benefits outweigh the costs.

The report of the Hinkley inquiry

Against this background, Michael Barnes produced his report of the public inquiry into the Hinkley Point C power station in the late summer of 1990. As the industry had hoped, he recommended that consent should be granted for the building of a new PWR. Despite this, his report was not an unqualified vindication of the industry's position. The economic viability of a new plant remained controversial, to be reconsidered as part of a governmental review of nuclear policy in 1994. Essentially, Barnes did little more than rule that Nuclear Electric (the successor to the CEGB's nuclear wing following the privatization of the non-nuclear part of the generating industry) should not be prevented in law from building the station if economic circumstances permitted. As he comments in his conclusions:

'If implementation by a private sector company is considered the present expectation is that the project would not prove financially attractive. The position therefore is that if consent is granted there is no certainty that it will be implemented.' (Barnes, 1990; Conclusions 8).

A possible ground for witholding consent in law would have been if Barnes had judged the risks to life to be intolerable. On this point, he accepted much of the argument put forward in the evidence from the HSE, and essentially adopted the standards proposed by them for the limits of tolerability. However, he introduced an important

189

point of principle affecting the definition of tolerability. He took the view that the statistical standards proposed (such as an annual risk of death of 1 in 100,000 for an individual member of the public) represented a level of danger above which the risks would be intolerable regardless of any benefits. If the risks were not as great as this, they might be either tolerable or intolerable, depending on perceived benefits; but without any benefits, even a very small risk would not be tolerable. The question of tolerability cannot thus be defined simply by the use of PRA as the application of the ALARP principle, as the HSE had proposed. It is a trade-off between a number of factors, and therefore much more fluid:

'It would be comforting if there had been some authoritative exposition of the maximum levels of individual and societal risk which could today be tolerated from the operation of a new nuclear power station in the light of some generally agreed benefits. Even if that exposition existed there would remain the disputed issue of what were the likely levels of the risk from a PWR at Hinkley Point. However, once the likely levels had been established the overall question would be decided. In fact no such exposition exists. Indeed, since the tolerable levels of risk must in logic be related to an assessment of benefits, and since the benefits can change, it is doubtful whether any long term levels of tolerable risk can ever be fixed.' (Barnes, 1990, 35.51).

Barnes further concluded that the effect of a new plant on the local area would be "generally disadvantageous". For example, while acknowledging some benefits due to employment and road improvement, he states:

'I would attribute more weight to the adverse impact ... on views from the Quantock Hills and from the coast to the west than to calculate risk of death to a person living a few miles from the plant of less than 1 in 100 million per year from the normal operation of the station. The adverse local effects have an added weight due to the fact that unlike other factors they are unlikely to change in any material way during the period of any consent which may be granted. Their assessment therefore contains an element of certainty not shared by some other factors.' (Barnes, 1990: Conclusions, 22).

190

The CEGB thus obtained the formal approval it was seeking for Hinkley Point C, but the report of the inquiry hardly amounts to a ringing endorsement of the case for nuclear power. It is as though the CEGB was being told 'There is no legal objection to you building a new power station at Hinkley Point but it is debatable whether it altogether a good idea for you to do so.' It is unclear whether the expression of such reservations by the inspector will make any difference to whether Hinkley Point C is ever built. Even so, an implicit precedent has been set with regard to the way in which opinions of the general public and local residents should be taken into account: ordinary people are not being 'irrational' if they object to nuclear developments in their neighbourhood. They have less access to, and understanding of, technical information on the basis of which quantitative risk assessments can be made, and so will generally be on weaker ground if the argument about risk comes down to numbers. However, the argument is about more than such numbers. If local residents do not stand to gain from the development, then (at least, according to Barnes, above a certain limit) whatever the numerical calculation of risk, such a risk will be intolerable. Furthermore, when the construction of a new power station will bring with it definable damage to the visual environment, we are left with a package which it is quite 'rational' for local communities to oppose. From the perspective of the industry, this is of course the classic NIMBY phenomenon, but to characterize local opposition as 'selfish' in this way is to imply a contradiction between local self-interest and some national interest. However, the case that nuclear power is required in the national interest still needs to be made out convincingly. Even if it is required, there remains a societal problem of inequity between local and national interest.

A future for nuclear power?

So if a new PWR is unlikely to be financially attractive and will be generally disadvantageous to the local area, is there anything to be said in its favour? Barnes supports a form of 'national interest' argument, by emphasizing the importance of retaining a nuclear option within the general strategy of the UK's national energy policy. Since the relative attractiveness (and price) of different energy options can vary with political and economic circumstances, it may be short-sighted to rule out technically feasible options just because they are presently too expensive. This principle is very relevant to broader structural changes in the British electricity industry, which is no longer a state-owned monopoly, but one in which privately owned suppliers supposedly compete with one another. Barnes is effectively saying that a privatized electricity industry operating in a free market cannot be trusted to produce a coherent long-term energy policy in the national interest if (as would be asserted by proponents of nuclear power) such a policy requires long-term investment. The profitability of nuclear power can be artificially enhanced by ignoring the costs of decommissioning of old plant. On the other hand, the capital costs of a new power station may look somewhat less daunting in the context of lower interest rates. We should not be mesmerized by financial forecasts any more than by statistical estimates of risk.

The second argument supported by Barnes appeals to an international interest. Since the burning of fossil fuels is a major cause of global warming, nuclear power (in its normal operation) may, paradoxically perhaps, be good for the world environment. Much more still needs to be understood about the causes of the greenhouse effect and the speed of its development. Even so, if the consequences of less nuclear power had been a conscious policy shift towards more burning of coal without a sufficiently compensating effort to make coal-fired stations less polluting, environmentalists would have scored a hollow victory. Here again there is the need to balance long-term and short-term consequences.

It would be reassuring to feel that such long-term environmental considerations were the driving force behind the planning of electricity generation for the early part of the next century. In the UK at least, this is far from being obviously the case. What matters is, above all, the price at which power can be supplied by the generators to the newly privatized electricity distribution companies. For the present, nuclear power is guaranteed a share of the market, but the logic of privatization requires that it should be

192

able to compete economically on equal terms, and it is far from clear how this can be achieved. As far as plans for new generating capacity are concerned, the power industry is being transformed by what has been called the 'rush to gas'. New gas-fired stations are relatively cheap and quick to build and operate, with few apparent ecological and economic risks when compared to nuclear. The story of the coal industry has been a much less happy one over the years since the start of our research, with a massive programme of pit closures wiping out the traditional industrial base of large areas of South Wales, and many parts of the Scottish Lowlands, Northern England and the English Midlands. One interpretation is that this will make what remains of the coal industry into a more attractive prospect for privatization.

Could nuclear power survive in a fully privatized environment? This is one of the main questions being addressed in the government's nuclear review, anticipated in the Barnes report, and now under way. In its submission to the review on June 20, 1994, Nuclear Electric has urged the government to press ahead with plans for its privatization so that it could have commercial freedom to compete with the non-nuclear privatized generating companies (National Power and PowerGen). However, it is clear that such 'commercial freedom' could only be attained if the taxpayer retained liability for those parts of the industry that no prudent investor would pay for. According to Nuclear Electric:

'The AGR stations and Sizewell B will be profitable without government subsidy. Risks over liabilities and uncertainties over income of these stations have been greatly reduced to the point where the majority of risks for the AGR and PWR stations can now be considered transferable to the private sector.'

However, the situation regarding the six remaining Magnox reactors is far more problematic, since:

'the risks associated with their back-end costs, resulting from decisions taken decades ago, would not be acceptable to the private sector.'

The 'back-end costs' referred to are those of decommissioning plants and removing the danger of radioactive leakage to the environment. Such costs were hardly considered in the early days of optimistic expansion, but they have since become a major headache for the industry worldwide. Privatization would thus require some legal device for insulating the company from these previous liabilities. If this were provided, what might follow? The new PWR at Sizewell B is nearing completion, and so:

> 'In order to exploit fully the national investment in Sizewell B, it is important to proceed with construction of a further plant as soon as possible.'

The first choice for such a 'further plant' would be a second PWR at Sizewell (Sizewell C), although the prospect of a (smaller and therefore cheaper) PWR at Hinkley Point C, instead or in addition, has not been abandoned. Could the necessary capital for the construction of such a new plant be obtained from the private sector? The argument seems to be coming down to percentages. A rate of return of 11 per cent might attract sufficient investment. Nuclear Electric claims to see a way of reaching something between 5 and 9 per cent, although this calculation is considered optimistic by both by the rival generating companies and environmentalist organizations (Financial Times, 21 June 1994). Would a subsidy from the government be politically acceptable and could it credibly close the gap? Whatever the outcome, it is clear that the nuclear industry needs to convince the market of its commercial viability, just as it needs to reassure the public about issues of health and safety.

Back ends and back yards

Even with no further expansion of nuclear generating capacity, however, many controversial policy decisions will still need to be made. We have mentioned the problems of making defunct reactors safe. The UK's six little Magnox stations could almost be dismissed as insignificant in a global context. Nations from the old Soviet bloc continue to depend on the power from the nuclear stations they have inherited, but require foreign capital to maintain anything approaching adequate standards of operational safety. Another extremely pressing problem is what to do with nuclear waste. At the time of our first study in this area (Eiser and Van der Pligt, 1979), reprocessing was the long-term policy most favoured by the UK government and the scientific establishment. Environmentalist opposition, such as it was, was against the principle of having a nuclear industry at all, rather than against the specific idea of reprocessing as opposed to storing waste. Approval for the Thermal Oxide Reprocessing Plant (THORP) at Sellafield was granted, not only on the basis of gross underestimates of construction time and cost, but on overestimates of eventual demand. THORP is at last now starting operation, but there is no programme of 'fast-breeder' reactors to use the plutonium produced by the reprocessing and no Soviet Union to offer an additional pretext for such investment on grounds of national security. Added to this are public fears over the international transportation both of the waste to be reprocessed and of the plutonium produced by the reprocessing. Despite optimistic noises from BNFL that THORP can still be profitable in relation to the costs of its operation, the costs of its construction appear to have been a massive lost investment of taxpayers' money.

But what else can be done? Whatever decisions could or should have been taken some twenty or more years ago, we are now left with consequences which cannot be simply wished away. Nuclear waste is accumulating, and much of it is being temporarily stored under unsatisfactory conditions. Despite the relatively early halt to expansion of the civil nuclear industry in the US (where no new orders for nuclear plants have been placed since 1978 with orders after 1974 eventually being cancelled), the problems of waste disposal there are as pressing as in many other parts of the world. According to a US Department of Energy (DOE) prediction in 1988, 41,000 tonnes of high-level radioactive waste will have accumulated by the year 2000, and 87,000 tonnes by the year 2020, even if no new reactors are built.

Dunlap, Kraft and Rosa (1993) report a considerable amount of survey data relating to the efforts of the DOE to find a publicly acceptable site for a High Level Nuclear Waste Repository

(HLNWR). The studies they describe are based on samples drawn from residents of communities close to the DOE's preferred site at Yucca Mountain, Nevada (90 miles north-west of Las Vegas) as well as at other sites originally considered. Their findings illustrate a widespread lack of trust among members of the general public in the nuclear industry and in official reassurances of nuclear safety. Although communities which have benefited economically in the past from the nuclear industry remain broadly positive, other communities tend to show an antipathy to all things nuclear. A combination of fears for health and safety and concerns about damage to other economic activities (especially tourism) make most residents opposed to having a repository in their own 'back yard', even if the need for such a repository somewhere is acknowledged in principle. A case could be made that the risks of radioactive pollution from an HLNWR are considerably less than from a new power station, but this does not appear to make such a lesser risk tolerable when there are insufficient compensatory benefits. Perceived inequity remains a powerful factor. The victories by local communities in keeping HLNWRs away from their doorsteps provide further proof, if any were needed, that the nuclear industry cannot expect to develop new facilities except over the objections of local residents. However, halting nuclear development of itself does not remove the problems already created. Waste is continuing to accumulate and much of it is being neither reprocessed nor safely stored in permanent repositories.

A future for attitude research?

The holding of an attitude involves interpreting facts in a particular way. It also involves attaching value to a particular interpretation. Clearly, people with different attitudes interpret and evaluate events in different ways. But what social psychological research also shows is how reluctant we often are to acknowledge the plausibility of viewpoints different from our own. Attitudes are not just 'matters of opinion' where we all happily agree to differ. They are claims about reality. If we say that nuclear power is really extremely safe (or dangerous), we are asserting this as a matter of fact, not just expressing a personal preference. We are, in other words, proposing a view of reality which we may need to defend against anyone who takes a contrary view. One way of mounting such a defence is to denigrate opposing points of view: this person's lack of concern is a sign of complacency, and we should not be reassured by it; this person's objection is based on ignorance and we should not be deterred by it.

For those faced with difficult technical or economic decisions, it must sometimes be very tempting to regard public opinion as a nuisance. When the public opinion in question may be not even that of the nation as a whole, but rather that of a few thousand villagers, the temptation for dismissing such troublesome objections must be even stronger. Moreover, it is easy to find psychological research which seems to justify giving in to this temptation. Ordinary people, by definition, have less expert knowledge. They can also be shown to be clumsy in their estimates of probability, to be unduly sensitive to particular kinds of publicity, and to be generally less concerned with other people's benefits and costs that with their own. With a small twist, this amounts to saying that the public is typically ill-informed, irrational, alarmist and selfish.

But even if this characterization of public opinion is correct, where does it leave policy-makers and industrial planners? Traditionally, it would appear that planners have accepted the need to make some concessions or compromises, no doubt, they would say, as the price of operating within a democratic system. These concessions, however, stop far short of surrendering the distinction between expert and non-expert judgment. For better or worse, policy decisions over matters such as nuclear power have to be made in a context where some account must be taken of what ordinary people think, but where the responsibility for such decisions is not simply handed over. This is a problem for nuclear power, precisely because such developments may be at least mildly unpopular, and perhaps extremely so. From the pronuclear perspective, the 'fact' of the matter is that nuclear power is extremely safe, but the public, 'mistakenly', believe it to be extremely dangerous. Why do the public feel this way? Because they are uninformed, and even constructively misinformed (by an unsympathetic popular press). What is to be done about this? One must find more effective ways of communicating the 'facts' about nuclear safety, so that the public will become better informed and therefore more pronuclear. What weight, in the meantime, does one give to verbal expressions of public opinion? Perhaps the apparent behavioural acceptance of many different kinds of risk in different contexts is a more dependable indicator of public concern.

All this looks just at one side of the equation. To accept or tolerate a risk one must believe that there are compensatory benefits. It could be that much (though maybe not all) public opposition to nuclear development arises from doubts over its supposed benefits and/or the possibility of better alternatives, rather than exclusively from fears about its safety. An understanding of public attitudes and hence of the disparity between public and industrial opinion must therefore look both at people's interpretation of risk, and at their interpretation of the balance of benefits and costs.

No means of electricity generation is completely without its costs or dangers, so there is a sense in which any choice of energy policy is a selection of the least unattractive of available solutions. However, these costs vary greatly and those of nuclear power are far from negligible, particularly if the risks of accidents are taken into account and particularly from the perspective of local residents. Some kind of case can still be made for nuclear power, even on environmental grounds. A case can be made for other energy options too, not least including conservation. Many of the arguments against nuclear power are not specifically antinuclear, but many of the arguments against burning fossil fuels are not specifically pronuclear either. Despite its potential for eliciting polarized opinions, the nuclear debate cannot be seen, nor decided upon, in isolation. There is a huge number of aspects that are potentially relevant, but ultimately some selection must be made among these aspects with regard to their relative importance if any decision is to be made. Certainty comes only from such selectivity, and dogmatism from a denial of the plausibility of priorities other than one's own.

Lessons from our research

The most important lesson from our research is that all attitudinal perspectives are relative. Rationality is a matter of making judgments according to a system of rules, but not all such rules are absolute. Some kinds of uncertainty are difficult to model, and some may matter much more than others. Assumptions have to be made before any formal analytic method can be applied to a decision problem, and these assumptions may ultimately be subjective. There is no single way of evaluating a given set of facts. What looks like a 'NIMBY' selfishness from one point of view can appear as justified self-interest — or, better still, a commitment to environmental protection — when viewed from the other side. As for the public being 'ill-informed', this depends on what one should be informed about in order to make a well-rounded decision. If terms such as 'alarmist' are introduced into the debate, others such as 'complacent' will surely follow.

We find clear evidence of such mutual denigration, for example in the responses of our Dorset residents to the post-Chernobyl questionnaire (Chapter 6), and in data from earlier research (Eiser and Van der Pligt, 1979). Polarization of opinion into opposing camps is also seen in the findings reported in Chapter 5, where we looked specifically at how the two sides in the debate characterized each other. If we put the pronuclear and antinuclear attitudinal perspectives alongside one another, we can see these perspectives for what they are — alternative and relative interpretations of reality. We do not have to say which interpretation is 'better', nor need we commit ourselves to saying that there is nothing in principle to choose between them. But it is still important to show the nature of the contrast, since it is from this contrast that disagreement stems. If instead we look at the issue purely from one perspective or another, we quickly lose sight of the distinction between reality and interpretation, between fact and value, and if we do this, we will never understand the divergences of attitude with which we are confronted. Our task as social psychologists is not merely to count the 'votes' on either side, but to examine and describe what it means to hold any one kind of attitude as distinct from any other.

199

A large part of what it means to have a given attitude is to have access to a particular store of knowledge and information. Attitudes are an expression of memories and learned associations, in other words, a reflection of many different kinds of experience. This principle leads us in a variety of complementary directions. Workers at a nuclear power station will know more about technical procedures than will members of the general public. Safety inspectors will be able to draw on experience from other industrial plants. Details of design, of maintenance and operating schedules, of supervision and training, will mean something to technologists in a way that they will not to most members of the general public. It is therefore only to be expected that the technological view of 'nuclear power' as a category will be more complex and differentiated than will be the view held by many ordinary people.

This ordinary, less differentiated, view is, as we have seen in Chapter 5, reflected in the standard treatment of energy issues in the media. We cannot say for sure how much of the knowledge-base of ordinary people's attitudes towards nuclear power is attributable to what they read in newspapers or see on television. However, we can say (from the findings reported in Chapter 6) that the public will generalize from particular pieces of 'bad news' about nuclear power and will form more negative attitudes towards nuclear developments that affect them more directly; moreover, it is those who are already more antinuclear who will be more prepared to see bad news from elsewhere as relevant to their local circumstances. From the pronuclear perspective, many such reactions to adverse publicity are attributable to overgeneralization, and to a neglect of crucial technical distinctions. On the other hand, there is no absolute answer to the question of the level at which events and objects should be categorized together or differentiated from one another. It is a fact that the design of a new PWR for Hinkley Point is technically quite different from the design of the infamous Chernobyl reactor. It is a matter of judgment whether these differences are sufficient — in the light partly of other factors which might transcend distinctions of technology — to allow us to categorize the Chernobyl accident as wholly irrelevant to issues of nuclear safety in Britain.

From this it follows that we cannot simply talk of people holding different attitudes on the same issue or towards the same object. Issues are represented differently at a psychological level by those whose attitudes are different. These differences in representation are reflected in the alternative systems of evaluation with which each side tends to label respective viewpoints. They are reflected, likewise, in the levels at which events are categorized and differentiated and in the accessibility of various memories, factual details, and associations. Additionally, they are reflected in differences in the salience or relative priority of particular aspects. This is possibly the most important message of our research. Differences in attitudes go along with differences in salience. This theme runs through many of our findings, particularly those reported in Chapters 3, 4 and 5. What strikes a supporter of nuclear power as most important about the proposal for a new power station will be very different from what seens important to an opponent. By documenting where these differences lie, social psychological research can help explain how and why supporters and opponents disagree.

But we can take this even further. Not everyone who objects to a local nuclear power station does so for the same reasons, nor do they do so in the same way. There are different forms of opposition, and our findings clearly document these. (Doubtless, too, there are different forms of support among more committed proponents, but this is a question which lies beyond the purposes and methods of the research we have reported.) Of special interest is the group who opposed a local development while being less opposed to nuclear development elsewhere. It is only too easy to characterize the views of such a group as reflecting a narrow 'NIMBY' egocentrism, or a 'parochial' disregard of the 'real' — i.e. broader national — issues. Our findings present a more sympathetic picture. Lots of things are 'real' and those things we can see and hear and touch for ourselves within a short walk from our own front doors deserve that epithet as surely as more remote projections of economic need or cancer incidence. Our 'local opponents' were anxious to protect their own personal quality of life. That much is undeniable. But in looking at what they regarded as most salient — at the constituents for them of such quality of life — we find an emphasis on conservation of the natural environment and the character of their neighbourhood, an emphasis which involves also the advocacy of a common good. If a different kind of development were proposed that still threatened such environmental goods, they would still oppose it. We see this in Chapter 4, in the forms of concern about oil-drilling by the Dorset coast, even though such development was regarded generally as far less threatening. The position of 'local opponents' is not antinuclear in principle, but this does not mean that it is unprincipled.

Attitudinal perspectives are relative in another sense as well. They imply comparisons. If 'local opponents' of a nuclear development ask the question 'Why do we have to have it here?', those who are more generally opposed ask 'Why does any development have to be nuclear?' This comes back to the salience of different aspects and forms of information. Renewable energy technologies tend to receive favourable media attention, and the more attention they receive, the more likely it seems to be that ordinary people will represent the issues of energy policy in terms that invite a choice between nuclear power and alternatives. Our simple manipulations of our questionnaire (reported in Chapter 4) show how powerful can be the effect — on certain very special kinds of response — of presenting more alternatives for consideration. Merely through wider discussion about diverse sources of energy, diversification of supply is likely to rise up the policy agenda. The perceived need for nuclear energy depends partly on what alternatives one can call to mind. What alternatives people actually call to mind will depend to a great extent on which they have been encouraged to think about. But while it is clear that many people would like more growth in reliance on renewable energy technologies, not all comparisons with other industries show nuclear power in a bad light. Coal-fired power stations are also viewed negatively, and the hypothetical prospect of a chemicals factory would be just as unacceptable to local residents as a new nuclear plant.

Our findings also suggest that the possibility of a future nuclear development is appraised rather differently by those who are familiar with the presence of an existing nuclear plant in their neighbourhood. Familiarity involves a different form of experience from which attitudes can be derived, and this is reflected in

202

differences in the salience of different aspects (see Chapter 3). As when considering the views of 'local opponents', self-interest cannot be totally disregarded as a factor, since nuclear power stations have a major economic impact on communities in their vicinity. Nor can we overlook the importance of self-selection: some people may choose to live near a nuclear plant (or at least feel unable or uninclined to move away) because of the employment and business opportunities it provides; in the same way others may choose to move to (or remain in) communities they expect to stay 'unspoilt'. Even so, our findings are far from being devoid of comfort as far as the nuclear industry is concerned. Misgivings about the desirability, necessity and safety of nuclear developments are evident in all our samples. However, such misgivings are least dominant or strongly expressed among those with most first-hand experience of living near a nuclear plant. So long as such first-hand experience (as was the case in our sample) is of a plant giving no obvious cause for alarm and creating no major emergency, it is likely to allay fears based on less immediate considerations.

Attitudes involve perspectives and perspectives depend on context. It is too easy to forget the importance of context when we think about issues with global implications. Whether to generate electricity through nuclear power or by other means is undoubtedly a global issue, although even here there will be different views about what kind of global issue it is. But it is also, always, a local issue, because any new power station has to be built somewhere. However 'global' our environmental or political concerns, we are all of us local residents of somewhere. In our research, we have tried to show how the attitudes of local residents deserve to be listened to seriously, in all their complexity, diversity and sincerity. These people are not, by and large, 'experts' in the normal sense of the word. In general, they are only too aware of how little they know of the technical aspects. But this does not mean that their views can be lightly set aside. In terms of the local context within which any proposed development could take shape, they are the ones with the first-hand knowledge. If we try to study attitudes while ignoring the context of real experience, we are left with the hollowness of empty words.

References

Abelson, R.P. (1959). Modes of resolution of belief dilemmas. *Journal of Conflict Resolution*, 3, 343-352.

Ajzen, I. & Fishbein, M. (1980). *Understanding attitudes and predicting social behavior*. Englewood Cliffs, N.J.: Prentice-Hall.

Allport, G.W. (1935). Attitudes. In: C. Murchison (Ed.) *Handbook of Social Psychology*. Worcester, Mass: Clark University Press.

Atom (1991). US Public expect a nuclear future. *ATOM*, 413, May 1991, p.3.

Barnes, M. (1990). *The Hinkley Point public inquiries*. London: HMSO.

Baum, A., Fleming, R., & Singer, J.E. (1982). Stress at Three Mile Island: applying social impact analysis. *Applied Social Psychology Annual*. Beverly Hills C.A.: Sage.

Black, D. (1984). *Investigation of the possible increased incidence of cancer in West Cumbria: Report of the independent advisory group*. London: Her Majesty's Stationery Office.

Bord, R.J. (1987). *Public cooperation as a social problem: the case of risky wastes*. Paper presented at the American Association for the Advancement of Science, 14-18 February 1987, Chicago, Illinois (9 pp.).

Breed, W. (1958). Social control in the newsroom: A functional analysis. *Social Forces*, 33, 326-335.

Brewer (1977). An information-processing approach to attribution of responsibility. *Journal of Experimental Social Psychology*, 13, 58-69.

Brown, J., Henderson, J. and Fielding, J. (1983). *Differing perspectives on nuclear related risks: on analysis of social psychological factors in the perception of nuclear power*. Paper presented at the meeting of the Operational Research Society, September, University of Warwick, England.

Burnett, P.C., Mann, L. & Beswick, G. (1988). *Validation of the Flinders Decision Making questionnaires in course decisions by students*. Unpublished manuscript. The Flinders University of South Australia.

Chapman, L.J. (1967). Illusory correlation in observational report. *Journal of Verbal Learning and Verbal Behavior*, 6, 151-155.

Cohen, B.L. and Lee, I.S. (1979). A catalog of risks. *Health Physics*, 36, 707-722.

Collins, D.L., Baum, A. and Singer, J.E. (1982) Coping with chronic stress at Three Mile Island: Psychological and biochemical evidence. *Health Psychology*, 2, 149-166.

205

Commission of the European Communities (1982). *Public opinion in the European Community* (Report No. XVII/202/83-E). Brussels, Belgium: Commission of the European Communities.

Corbett, J. (1987). The safety of nuclear power in a risky world. *Atom, 368* (June), 20-23.

Craft, A.W., Openshaw, S. & Birch, J.M. (1984). Apparent clusters of childhood lymphoid malignancy in Northern England. *The Lancet,* July 14th, 2, 96-97.

Crano, W.D. (1983). Assumed consensus of attitudes: The effect of vested interest. *Personality and Social Psychology Bulletin, 9,* 597-608.

Cunningham, S. (1985). The public and nuclear power. American Psychological Association, *Monitor, 16*: 1.

Davidson, L.M., Baum, A. and Collins, D.L. (1982). Stress and control-related problems at Three Mile Island. *Journal of Applied Social Psychology, 12,* 349-359.

Davidson, L.M., Baum, A., Fleming, I., & Gisriel, M.M. (1986). Toxic exposure and chronic stress at Three Mile Island In: A.H. Lebovits, A. Baum, & J.E. Singer (Eds.) *Advances in Environmental Psychology, Vol. 6,* pp. 35-46. Hillsdale NJ: Erlbaum.

Decima (1987). *A study of Canadians' attitudes toward the use of nuclear energy to generate electricity in Canada. Report # 2383.* Decima Research, Ontario, Canada.

Dohrenwend, B.P., Dohrenwend, B.S., Kasl, S.V. and Warheit, G.J. (1979). *Report of the Task Group on Behavioral Effects to the President's Commission on the accident at Three Mile Island.* Washington, D.C.: USA Government Printing Office.

Dunlap, R.E., Kraft, M.E., & Rosa, E.A. (eds.) (1993). *Public reactions to nuclear waste: Citizens' views of repository siting.* Durham, NC: Duke University Press.

Dupont, R.L. (1980). Nuclear phobia: phobic thinking about nuclear power. In: *Nuclear Power in American thought,* (23-41), Washington D.C.: Edison Electric Institute.

Earle, T.C. & Cvetkovich, G. (1990). What was the meaning of Chernobyl? *Journal of Environmental Psychology, 10,* 169-176.

Edwards, W. (1954). The theory of decision making. *Psychological Bulletin, 51,* 380-417.

Eiser, J.R. (1990). *Social judgment.* Buckingham: Open Univeristy Press.

Eiser, J.R., ed. (1984). Attitudinal judgment. New York: Springer-Verlag.

Eiser, J.R. & Van der Pligt, J. (1982). Accentuation and perspective in attitudinal judgement. *Judgement of Personality and Social Psychology, 42*, 224-238.

Eiser, J.R. & Van der Pligt, J. (1988). *Attitudes and decisions.* London: Routledge.

Eiser, J.R. and Van der Pligt, J. (1979). Beliefs and values in the nuclear debate. *Journal of Applied Social Psychology, 9*, 524-536.

Eiser, J.R., Van der Pligt, J., & Spears, R. (1989)Local residents attributions for nuclear decisions. *Basic and Applied Social Psychology, 10*, 141-148.

Erikson, K.T. (1990). Toxic reckoning: Business faces a new kind of fear. *Harvards Business Review* (January-February); 118-126.

Ester, P., Mindell, C., Van der Linden, J. and Van der Pligt, J. (1983). The infuence of living near a nuclear power plant on beliefs about nuclear energy. *Zeitschrift für Umweltpolitik (Journal of Environmental Policy), 6*, 349-362.

Farhar-Pilgrim, B. & Freudenburg, W.R. (1984). Nuclear energy in perspective: A comparative assessment of the public view. In W.R. Freudenberg & E.A. Rosa (Eds.) *Public reaction to nuclear power: Are ther critivcal masses?* Westview Press: Colorado.

Festinger, L. (1957). *A theory of cognitive dissonance.* New York: Harper and Row.

Fischhoff, B. (1983) Predicting frames. *Journal of Experimental Psychology: Learning, Memory and Cognition, 9*, 103-116.

Fischhoff, B., Lichtenstein, S., Slovic, P., Derby, S.L., & Keeney, R.L. (1981). *Acceptable risk.* Cambridge: Cambridge University Press.

Fishbein, M. (1967). Attitude and the prediction of behavior. In M. Fishbein (ed.) *Readings in attitude theory and measurement.* New York: Waley. pp 477-492.

Fishbein, M. and Ajzen, I. (1975). *Belief, attitude, intention and behavior.* Reading, Mass.: Addison Wesley.

Fiske, S.T., & Taylor, S.E. (1991). *Social cognition.* (2nd Ed.) New York: Random House.

Flynn, C.B. (1979). *Three Mile Island Telephone Survey: A perliminary report.* Washington D.C.: US Nuclear Regulatory Commission.

Flynn, C.B. (1981). Local public opinion. In: T.H. Moss and D.L. Sills (Eds.). The Three Mile Island nuclear accident: Lessons and implications. *Annuals of the New York Academy of Sciences, Vol 365*, 146-158. New York: New York Academy of Sciences.

Frankena, F. (1983). Facts, values and technical expertise in a renewable energy siting dispute. *Journal of Economic Psychology, 4*, 131-147.

Freudenburg, W.R. and Baxter, R.K. (1984). Nuclear reactions: Public attitudes and policies toward nuclear power. *Policy Studies Review, 5*, 96-110.

Freudenburg, W.R. and Baxter, R.K. (1984). Host community attitudes toward nuclear power plants: A reassessment. *Social Science Quarterly, 65*, 1129-1134.

Gardner, M.J., Snee, M.P., Hall, A.J., Powell, C.A., Downes, S. & Terrell, J.D. (1990). Results of a case-control study of leukaemia and lymphoma among young people near Sellafield nuclear plant in West Cumbria. *British Medical Journal*, February 17th, *300* (6722), 423-429.

Gerbner, G., & Gross, L. (1976). Living with television: The violence profile. *Journal of communication, 26*, 173-199.

Gerbner, G., Gross, L., Morgan, M., & Signorelli, N. (1980). The 'mainstreaming' of America: Violence profile no. 11. *Journal of communication, 30*, 10-29.

Gittus, J.H. (1987). The Chernobyl accident and its consequences. *Atom, 368* (June), 2-9.

Hamilton, D.L. & Gifford, R.K. (1976). Illusory correlation in interpersonal perception: A cognitive basis of stereotypic judgments. *Journal of Experimental Social Psychology, 12*, 392-407.

Hamilton, D.L. & Rose, T.L. (1980). Illusory correlation and the maintenance of stereotypic beliefs. *Journal of Personality and Social Psychology, 39*, 832-845.

Hogarth, R.M. (1981) Beyond discrete biases: functional and dysfunctional aspects of judgmental heuristics. *Psychological Bulletin, 90*, 197-217.

Hovland, C.I. & Sherif, M. (1952). Judgmental phenomenon and scales of attitude measurement: Item displacement in Thurstone Scales. *Journal of Abnormal and Social Psychology, 47*, 822-832.

Hughey, J.B., Sundstrom, E. and Lounsbury, J.W. (1985). Attitudes toward nuclear power: a longitudinal analysis of expectancy-value models". *Basic and Applied Social Psychology, 6*, 75-91.

Janis, I.L. & Mann, L. (1977). *Decision making: A psychological analysis of conflict, choice, and commitment.* New York: Free Press.

Johnson, B.B. (1987). Public concerns and the public role in siting nuclear and chemical waste facilities. Environmental Management, 11, 571-586.

Judd, C.M., & Johnson, J.T. (1981). Attitudes, polarization and diagnosticity: Exploring the effect of affect. *Journal of Personality and Social Psychology, 41*, 26-36.

208

Judd, C.M., & Johnson, J.T. (1984). The polarizing effects of affective intensity. In J.R. Eiser (ed.) *Attitudinal judgment.* New York: Springer.

Kahneman, D. & Tversky, A. (1979). Prospect theory: An analysis of decision under risk. *Econometrica, 47,* 263-291.

Kasperson, R.E. (1980). The dark side of the radioactive waste problem. In: T. O'Riordon and K. Turner (Eds.) *Progress in Resource Management and Environmental Planning, Vol. 2,* 133-163. Chichester: John Wiley.

Kasperson, R.E. (1985). *Rethinking the siting of hazardous waste facilities.* Paper presented at the conference on Transport, Storage, and Disposal of Hazardous Materials. IIASA, Vienna, Austria, July 1985.

Kasperson, R.E., Berk, G., Pijawka, D., Sharaf, A.B. & Wood, J. (1980). Public opposition to nuclear energy: Retrospect and prospect. *Science, Technology and Human Values, 5,* 11-23.

Kates, R.W. and Braine, B. (1983). Locus, equity, and the West Valley nuclear wastes. In: R.E. Kasperson and M. Berberian (Eds.) *Equity issues in radioactive wastemanagement,* 301-331. Cambridge, Mass.: Oelgeschlager, Gunn and Hain.

Katz, E. & Lazarsfeld, P.F. (1957). *Personal influence.* Glencoe, Illinois: Free Press.

Katz, E. (1957). The two-step flow of communication: An up-to-date report on a hypothesis. *Public Opinion Quarterly, 21,* 61-78.

Kraybill, D.B. (1979). *Three Mile Island*: Local residents speak out. Unpublished reports Social Science Center. Elizabethtown College, Elszabethtown, Penn.

Layfield, F. (1987). *Sizewell B public inquiry.* London: HMSO.

Lemert, J.B. (1981). *Does mass communication change public opinion after all? A new approach to effects analysis.* Chicago: Nelson-Hall.

Lemert, J.B., Mitzman, B.N., B.N., Seither, M.A., Cook, R.M., & Hackett, R. (1977). Journalists and mobilizing information. *Journalism Quarterly, 54,* 721-726.

Lichtenstein, S., Slovic, P., Fischhoff, B., Layman, N., & Coombs, B. (1978). Judged frequency of lethal events. *Journal of Experimental Psychology: Human Learning and Memory, 4,* 551-578.

Lindell, M.K. & Perry, R.W. (1990). Effects of the Chernobyl accident on public perceptions of nuclear plant accident risks. *Risk Analysis, 10,* 393-400.

Lowe, P., & Morrison, D. (1984). Bad news or good news: Environmental politics and the mass media. *The Sociological Review, 32,* 75-90.

Lowry, S. & DeFleur, M.L. (1983). *Milestones in mass communication research: Media effects.* London: Longman.

MacKuen, M.B. & Coombs, S.L. (1981). *More than news: Media power in public affairs*. Sage: Beverly Hills.

Mann, L. (1982). *Decision Making Questionnaires I and II*. Unpublished Scales. Flinders Decision Workshops, The Flinders University of South Australia.

Marks, G. & Miller, N. (1987). Ten years of research on the false-consensus effect: An empirical and theoretical review. *Psychological Bulletin, 102*, 72-96.

Mazur, A. (1984). Media influences on public attitudes toward nuclear power. In W.R. Freudenberg & E.A. Rosa (Eds.) *Public reaction to nuclear power: Are ther critivcal masses?* Westview Press: Colorado.

Mazur, A. (1990). Nuclear Power, Chemical hazards, and the quality-of-reporting theory of media effects. *Minerva, 28*, 294-323.

McCombs, M.E. & Shaw, D.L. (1972). The agenda-setting function of mass media. *Public Opinion Quarterly, 36*, 176-187.

McGuire, W.J. (1986). The myth of massive media impact: Savagings and salvagings. In G. Gomstock (Ed.) *Public communication and behavior*. Orlando: Academic Press.

McLeod, J.M., Becker, L.B., & Byrnes, J.E. (1974). Another look at the agenda-setting function of the press. *Communication Research, 1*, 131-166.

McQuail, D. (1987). *Mass communication theory: An introduction*. London: Sage.

Melber, B.D., Nealey, S.M., Hammersla, J. and Rankin, W.L. (1977). *Nuclear Power and the public: Analysis of collected survey research.* PNL-2430 Seattle: Battelle Human Affairs Research Centers.

Meyerowitz, B.E. & Chaiken, S. (1987). The effect of message framing on breast self-examination: Attitudes, intention and behavior. *Journal of Personality and Social Psychology, 52*, 500-510.

Midden, C.J.H. and Verplanken, B. (1986). *Na Tsjernobyl ... Enige conclusies over effecten van het ongeluk in Tsjernobyl op de publieke opinie over kernenergie*. Petten: ECN Report ESC-WR-86-23.

Midden, C.J.H. and Verplanken, B. (1990). The stability of nuclear attitudes after Chernobyl. *Journal of Environmental Psychology, 10*, 111-120.

Mitchell, R.C. (1980). *Final results of the resources for the future national environment survey for the president's Council on Environmental Quality*. Washington, D.C.: Resources for the Future.

Mitchell, R.C. (1984). Rationality and irrationality in the public's perception of nuclear power. In Freudenberg, W.R. and Rosa, E.A. (Eds.) *Public reactions to nuclear power: Are there critical masses?* 137-179. Boulder, Colorado: Westview Press

Mullen, B., & Hu, L. (1988). Social projection as a function of cognitive mechanisms: Two meta-analytic integrations. *British Journal of Social Psychology, 27,* 333-356.

Mullen, B., Atkins, J.L., Champion, D.S., Edwards, C., Hardy, D., Story, J. E., & Vanderklok, M. (1985). The false consensus effect: A meta-analysis of 115 hypothesis tests. *Journal of Experimental Social Psychology, 21,* 262-283.

Murdock, G. (1974). Mass communication and the construction of meaning. In N. Armistead (Ed.) *Reconstructing social psychology.* Harmondsworth: Penguin.

Nealey, S.M., Melber, B.D. and Rankin, W.L. (1983). *Public Opinion and nuclear energy.* Lexington, M.A.: Lexington.

Nisbett, R.E., & Ross, L. (1980). *Human inferences: strategies and short comings of social judgment.* Englewood Cliffs, N.J.: Prentice Hall.

Nucleonic Week (1986). Antinuclear fallout from Chernobyl continues to wash over Europe. *Nucleonic Week,* May 1986, 11-13.

O'Riordan, T., Kemp, R. & Purdue, H.M. (1988). *Sisewell B: An anatomy of the inquiry.* London: Macmillan.

Otway, H., Maurer, D., and Thomas, K. (1978). Nuclear power: The question of public acceptance. *Futures, 10,* 109-118.

Otway, H.J. and Fishbein, M. (1976). *The determinants of attitude formation: An application to nuclear power.* (Research Memorandum RM76-80). Laxenburg, Austria: International Institute for Applied Systems Analysis.

Pahner, P.D. (1976). *A psychological perspective of the nuclear energy controversy.* International Institute for Applied Systems Analysis Report *RM-76-67.* Laxenburg, Austria.

Paletz D.L., Reichart, P., & McIntyre, B. (1971). How the media support local governement authority. *Public Opinion Quarterly, 35,* 80-92.

Peltu, M. (1985). The role of communications media. In H. Otway & M. Peltu (Eds.) *Regulating industrial risks,* London: Butterworths.

Petty, R.E. & Cacioppo, J.T. (1986). *Communication and persuasion: Central and peripheral routes to attitude change.* New York: Springer.

Popper, F.J. (1981). Siting LULU's. *Planning* (April): 12-15.

Radford, M., Mann, L. & Kalucy, R. (1986). Psychiatric disturbance and decision making. *Australian and New Zealand Journal of Psychiatry, 20,* 210-217.

Rankin, W.L. and Nealey, S.M. (1978). Attitudes of the public about nuclear wastes. *Nuclear News, 21,* 112-117.

211

Rankin, W.L., Melber, B.D., Overcast, T.D. & Nealy, S.M. (1981). *Nuclear power and the public: An update of collected survey research on nuclear power.* Seattle: Batelle Human Affairs Research Centers.

Rasmussen, N.C. (1975). *Reactor safety study: An assessment of accident risks in U.S. commercial nuclear power plants.* U.S. Nuclear Regulatory Commission, WASH-1400.

Reason, J.T. (1990). *Human error.* Cambridge: Cambridge University Press.

Renn, O. (1990). Public responses after Chernobyl: Effects on attitudes and public policies. *Journal of Environmental Psychology, 10,* 151-168.

Renn, O. (1990). Public responses to the Chernobyl accident. *Journal of Environmental Psychology, 10,* 151-167.

Roberts, D.F. & Maccoby, N. (1985). Effects of mass communication. In G. Lindzey & E. Aronson (Eds.) *The handbook of social psychology (3rd Ed.) Vol. 2.* New York: Random House.

Rosa, E.A., & Freudenburg, W.R. (1993). The historical development of public reactions to nuclear power: Implications for nuclear waste policy. In: R.E. Dunlap, M.E. Kraft, & E.A. Rosa (Eds.) *Public reations to nuclear waste: Citizens' views of repository siting.* pp. 33-63. Durham & London: Duke University Press.

Rosenberg, M.J. and Hovland, C.I. (1960). Cognitive, affective and behavioral components of attitudes. In: M.J. Rosenberg, C.I. Hovland, W.J. McGuire, R.P. Abelson and J.W. Brehm (Eds.) *Attitude organization and change: An analysis of consistency among attitude components,* 1-14. New Haven, Conn: Yale University Press.

Ross, L., Greene, D., & House, P. (1977). The 'false consensus effect': An egocentric bias in social perception and attribution processes. *Journal of Experimental Social Psychology, 13,* 279-301.

Saunders, P. (1987). Nuclear benefits and risks. *Atom, 365* (March), 20-21.

Schwarz, N., Hippler, H.J., Deutsch, B. & Strack, F. (1985). Response scales: Effects of category range on reported behavior and comparative judgments. *Public Opinion Quarterly, 49,* 388-395.

Seligman, M.E.P. (1975). *Helplessness.* San Francisco: Freeman.

Sherif, M. & Hovland, C.I. (1961). *Social Judgment: Assimiliation and contrast effects in communication and attitude change.* New Haven, Conn.: Yale University Press.

Slovic, P., Fischhoff, B., & Lichtenstein, S. (1982). Facts versus fears: Understanding perceived risk. In Kahneman, D., Slovic, P., & Tversky, A. (Eds.) *Judgment under uncertainty: Heuristics and biases.* Cambridge: CUP.

Soderstrom, E.J., Sorensen, J.H., Copenhaver, E.D. and Carnes, S.A. (1984). Risk perception in an interest group context: An examination of the TMI restart Issue. *Risk Analysis, 4,* 231-244.

Spears, R., & Manstead, A.S.R.M. (1990). Consensus estimation in social context. In W. Stroebe & M. Hewstone (Eds.) *European Review of Social Psychology, Vol. 1.* Chichester: Wiley.

Spears, R., Eiser, J.R. & Van der Pligt, J. (1987). Further evidence for expectation-based illusory correlations. *European Journal of Social Psychology, 17,* 253-258.

Spears, R., Eiser, J.R. & Van der Pligt, J. (1989). Attitude strength and perceived prevalence of attitude postions. *Basic and Applied Social Psychology, 10,* 43-55.

Spears, R., Van der Pligt, J., & Eiser, J.R. (1985). Illusory correlation in the perception of group attitudes. *Journal of Personality and Social Psychology, 48,* 863-875.

Spears, R., Van der Pligt, J., & Eiser, J.R. (1986a). Evaluation of nuclear power and renewable alternatives as portrayed in UK local press coverage. *Environment and Planning A, 18,* 1629-1647.

Spears, R., Van der Pligt, J., & Eiser, J.R. (1986b). Generalizing the illusory correlation effect. *Journal of Personality and Social Psychology, 51,* 1127-1134.

Spears, R., Van der Pligt, J., & Eiser, J.R. (1987). Sources of evaluation of nuclear and renewable energy contained in the local press. *Journal of Environmental Psychology, 7,* 310-43.

Starr, C. (1969). Social benefit versus technological risk. *Science, 165,* 232-238.

Sundstrom, E., Lounsbury, J.W., De Vault, R.C. and Peele, E. (1981). Acceptance of a nuclear power plant: Applications of the expectancy value model. In. A. Baum and J.E. Singer (Eds.), *Advances in environmental psychology, Vol. 3,* 171-189. Hillsdale, NJ: Lawrence Erlbaum Associates, Inc.

Swets, J.A. (1973). The receiver operating characteristic in psychology. *Science, 182,* 990-1000.

Tesser, A. (1978). Self-generated attitude change. In L. Berkowitz (Ed.). *Advances in Experimental Social Psychology. (Vol 11).* New York: Academic press.

Thomas, K. and Baillie, A. (1982). *Public attitudes to the risks, costs, and benefits of nuclear power.* Paper presented at a joint SERÇ/SSRC seminar on research into nuclear power development policies in Britain, June, 1982.

213

Thomas, K., Maurer, D., Fishbein, M., Otway, H.J., Hinkle, R., & Simpson, D.(1980). A comparative study of public beliefs about five energy systems. RR-80-15. International Institute for Applied Systems Analysis, Laxenburg, Austria.

Turner, J.C. Hogg, M.A., Oakes, P.J., Reicher, S.D., & Wetherell, M. (1987). *Rediscovering the social group: A self-categorization theory.* Oxford: Blackwell.

Tversky, A. & Kahneman, D. (1973). Auxilability: a heuristic for judging frequency and probability. *Cognitive Psychology, 5,* 207-232.

Tversky, A. & Kahneman, D. (1974). Judgment under uncertainty: heuristics and biases. *Science, 185,* 1127-1131.

Tyler, T.R. & Cook, F.L. (1984). The mass media and judgments of risk: Distinquishing impact on personal level and societal level judgments. *Journal of Personality and Social Psychology, 47,* 693-708.

Van der Pligt, Eiser, J.R. and Spears, R. (1986b). Attitudes toward nuclear energy: Familiarity and Salience, *Environment and Behaviour, 18,* 75-93.

Van der Pligt, J., Eiser, J.R. and Spears, R. (1986a). Construction of a nuclear power station in one's locality: Attitudes and Salience. *Basic and Applied Social Psychology, 7,* 1-15.

Van der Pligt, J., Eiser, J.R. and Spears, R. (1987a). Comparative judgments and preferences: The influence of the number of response alternatives. *British Journal of Social Psychology, 26,* 269-280.

Van der Pligt, J., Eiser, J.R. and Spears, R. (1987b). Nuclear waste: Facts, fears and attitudes. *Journal of Applied Social Psychology, 17,* 453-470.

Van der Pligt, J., Van der Linden, J. and Ester, P. (1982). Attitudes to nuclear energy: Beliefs, values and false consensus. *Journal of Environmental Psychology, 2,* 221-231.

Van der Pligt, J. (1992). *Nuclear Energy and the Public.* Oxford: Blackwell.

Van der Pligt., Eiser, J.R. and Spears, R. (1987a). Comparative judgments and preferences: The influence of the number of response alternatives. *British Journal of Social Psychology, 26,* 269-280.

Van Schie, E.C.M., & Van der Pligt, J. (1994). Getting an anchor on availability in causal judgment. *Organizational Behavior and Human Decision Processes, 57,* 140-154.

Warren, D.S. (1981). Local attitudes to the proposed Sizewell "B" nuclear reactor. *Report RE19,* Food and Energy Research Centre, October 1981.

Webley, P. & Spears, R. (1986). Economic preferences and inflationary expectations. *Journal of Economic Psychology, 7,* 359-369.

214

Webley, P., Eiser, J.R. & Spears, R. (1988). Inflationary expectations and policy preferences. *Economics Letters, 7,* 359-369.

Woo, T.O. and Castore, C.H. (1980). Expectancy-value and selective exposure determinants of attitudes toward a nuclear power plant. *Journal of Applied Social Psychology, 10,* 224-234.

Zanna, M.P., Olson, J.M., & Herman, C.P. (Eds.) (1987). *Social influence: The Ontario Symposium, Vol. 5.* Hillsdale: Erlbaum.

Subject Index

217

Author Index